AF870098

STP 1019

Novel Techniques in Fossil Fuel Mass Spectrometry

Terrence R. Ashe and Karl V. Wood, editors

ASTM
1916 Race Street
Philadelphia, Pa. 19103

Library of Congress Cataloging-in-Publication Data

Novel techniques in fossil fuel mass spectrometry/Terrence R. Ashe and Karl V. Wood, editors
(STP; 1019)
Papers presented at a symposium held May 24–29, 1987, in Denver, Colo., and sponsored by the ASTM Committee E-14 on Mass Spectrometry.
"ASTM publication code number 04-010190-50"—T.p. verso.
Includes bibliographical references and index.
ISBN 0-8031-1198-3
1. Fuel—Analysis—Congresses. 2. Mass spectrometry—Congresses. I. Ashe, Terrence R. II. Wood, Karl Vernon. III. ASTM Committee E-14 on Mass Spectrometry. IV. Series: ASTM special technical publication; 1019.
TP321.N68 1989
662′.6—dc19 89-352
CIP

NOTE

The Society is not responsible, as a body, for the statements and opinions advanced in this publication.

Peer Review Policy

Each paper published in this volume was evaluated by three peer reviewers. The authors addressed all of the reviewers' comments to the satisfaction of both the technical editor(s) and the ASTM Committee on Publications.

The quality of the papers in this publication reflects not only the obvious efforts of the authors and the technical editor(s), but also the work of these peer reviewers. The ASTM Committee on Publications acknowledges with appreciation their dedication and contribution of time and effort on behalf of ASTM.

Printed in Baltimore, MD
April 1989

Foreword

This publication, *Novel Techniques in Fossil Fuel Mass Spectrometry,* contains papers presented at a symposium on Fossil Fuel Analysed by Mass Spectrometry which was held in Denver, Colorado, 24–29 May 1987. The symposium was sponsored by ASTM Committee E-14 on Mass Spectrometry. Terrence R. Ashe, Esso Petroleum Canada, and Karl V. Wood, Purdue University, presided as symposium chairmen and are editors of this publication. The American Society of Mass Spectroscopy handled all the arrangements for the symposium.

Contents

Overview

In recent years new ionization techniques, including laser desorption, fast atom bombardment and plasma desorption, and new separation techniques, combined with mass spectrometry, including tandem mass spectrometry, liquid chromatography, and supercritical fluid chromatography, have become standard analytical mass spectrometric capabilities. This book describes recent developments in the applications of these new techniques, including some computer software innovations, to the complex problems associated with the fossil fuel industry. These techniques bring new mass spectrometric capabilities to the fossil fuel analytical laboratory to complement the standard capabilities already in use, like low- and high-resolution electron ionization, chemical ionization, and gas chromatography/mass spectrometry. This book describes the details for better understanding some of these novel techniques and the application of given techniques to the solution of specific problems. Therefore, this book will provide the practicing researcher with a better appreciation for the application of specific techniques in fossil fuels analysis. One good example is the increasing need to analyze higher molecular weight, nonvolatile polyfunctional materials (such as resids) without thermally degrading them. In time, these novel ionization and separation techniques will have a dramatic impact on the fossil fuels analytical chemist.

Background

Fast atom bombardment (FAB) involves directing a neutral atom beam (argon, in most cases) at a liquid matrix in which the sample is dissolved. The liquid matrix is typically glycerol due to its relatively low vapor pressure. The ions observed can result from precharged species or protonation of neutral molecules. FAB has been shown to give molecular weight information for a wide range of chemical classes. In addition to glycerol, a large array of matrices have been used including thioglycerol, dithiothreitol, dithioerythritol, lithium sulfolane, and dimethylsulfoxide to name a few of the most common solvents. Campana and Freas[1] have utilized both positive and negative FAB for the screening of additives in base oils and lubricants. While negative ion FAB provided less structural information it proved more useful as a means for profiling the various materials. An ionization technique closely related to FAB is organic secondary ion mass spectrometry (SIMS) in which an ion beam is used instead of the neutral atom beam of FAB. In SIMS either a liquid matrix or a dry sample can be used.

Another relatively new ionization technique for obtaining molecular weight information from nonvolatile and thermally labile compounds is laser desorption. In this technique the sample is ionized by photon irradiation. Both continuous wave (for example, carbon dioxide) and pulsed (for example, neodymium: yatrium aluminum garnet) lasers have been used. Data acquisition is an important parameter for laser desorption due to the time

[1] Campana, J. E. and Freas, R. B., *Transactions,* American Society of Lubrication Engineers, Vol. 29, No. 2, 1986, p. 235.

dependence of ion formation and subsequent collection. For example, Brown and Wilkins[2] have utilized a pulsed CO_2 laser coupled to a Fourier Transform Mass Spectrometer for laser desorption studies of synthetic porphyrins. The $M^{+\cdot}$ ions were favored by high laser energy output whereas the cationized species were favored at low laser energy output.

The final novel ionization technique to be discussed in this brief review is plasma desorption ionization. Typically this technique utilizes ^{252}Californium as the ionizing source, such that the nuclear event causes atom fission with one fragment starting a clock and the other penetrating the sample with subsequent ionization. The resulting ions are then mass analyzed with a time-of-flight mass spectrometer.

Tandem mass spectrometry (MS/MS) a relatively new separation technique for the direct analysis of complex mixtures, is comprised of two mass analyzers separated by a collision cell. Analogous to gas chromatography/mass spectrometry (GC/MS) which separates neutrals prior to ionization and mass analysis, MS/MS separates by mass-to-charge ratio with the first mass analyzer (after the mixture has been ionized), passing the separated ions into the collision cell where they undergo collision induced dissociation (CID). The resulting fragments are then mass analyzed with the second mass analyzer. In this manner the CID spectra of individual mass-to-charge ions (termed daughter spectra) can be obtained. The spectra can then be compared with authentics or interpreted a priori in a similar manner to GC/MS data. In addition to this mode of operation other scan modes are also amenable to MS/MS increasing the selectivity for specific compound types (for example, parent scans for characterizing a complex mixture for those components having a selected substructure or neutral loss scans for characterizing a complex mixture for those components having a specific functional group). A number of fuel-related studies in which tandem mass spectrometry was utilized have been reported. For example, Wood et al.[3] have shown that the combination of a calcium/mixed amine reduction of an SRC II middle distilled fraction with MS/MS parent scans proved invaluable for characterizing this complex coal-derived liquid for sulfur-containing polynuclear aromatic hydrocarbons (PNAs). Reduction of the benzothiophene and dibenzothiophene-type molecules resulted in C-S bond scission with subsequent hydrogenation of the nonthiophenolic portion of the molecule. Using negative ion chemical ionization parent scans for the negative thiatropylium ion enabled characterization of the entire series of alkyl benzothiophenes and dibenzothiophenes.

Another novel mass spectrometric separation technique involves the coupling of liquid chromatography to the mass spectrometer. This technique enables nonvolatile thermally labile compounds to be successfully analyzed with a mass spectrometer. The major breakthrough enabling this successful coupling involved the interface—one of the most successful being a combination interface/ionization assembly called thermospray. In this interface the LC effluent is utilized as a medium for ion formation. However, solvents, salts, solvent flow rate, and interface temperature are all important parameters for successful ion formation using the thermospray interface.

The most recent separation technique combined with mass spectrometry is supercritical fluid chromatography. Supercritical fluid chromatography has been shown to have some enhanced separation potential which surpasses both gas chromatography and liquid chromatography. This technique utilizes the fact that the properties of the supercritical fluid (that is, low viscosity) result in the mobile phase having a character intermediate between

[2] Brown, R. S. and Wilkins, C. L., *Analytical Chemistry,* Vol. 58, 1986, p. 3196.

[3] Wood, K. V., Cooks, R. G., Laugal, J. A., and Benkeser, R. A., *Analytical Chemistry,* Vol. 57, 1985, p. 692.

a gas and a liquid. While CO_2 is a common supercritical mobile phase the addition of other more polar solvents increases the separating capability.

Highlights

The papers in this book can be divided roughly into three general categories. The first few papers deal with some relatively new approaches to fossil fuel analysis using mass spectrometry. The majority of papers, comprising the middle section, apply novel techniques to specific problems. The last two papers deal with the impact computers have in coupling mass spectrometry with real world analyses.

The three relatively new approaches to fossil fuel analysis presented include supercritical fluid chromatography/mass spectrometry, GC/MS hydrocarbon type analysis and chemical ionization proton exchange analysis. Chess et al. detail the development of supercritical fluid chromatography (SFC)/mass spectrometry for fossil fuels analysis. Particular emphasis is placed on both the use of polar fluids and the interfaces available for doing SFC. Furthermore they show examples of chemical class separation using SFC/MS. They feel that the advancement of SFC/MS in fossil fuels analysis is limited, at present, by the number of active participants, but the potential for major contributions using SFC/MS is there. Gehran and Yost describe development of a method for using GC/MS to do hydrocarbon type analysis (HTA) of jet fuels. This method utilizes a short DB5 capillary column for sample introduction. The type analysis generated is based on the current ASTM D 2789 inverse calibration matrices. Particular emphasis was placed on investigating component discrimination in GC introduction as well as sensitivity and limits of detection. The GC/MS/HTA results were found to be both reproducible and relatively consistent with ASTM D 2789 results which represent the industry standard. Miller and Hawthorne describe the use of chemical ionization—proton exchange (CIPE) for the determination of the number of unsubstituted positions on an aromatic molecule. This information can be utilized to identify a number of isomeric aromatic species whose electron ionization mass spectra are indistinguishable. This method utilizes methanol-d_1 for the reagent gas. Upon ionization only the aromatic hydrocarbons undergo deuterium exchange. CIPE analysis was shown to be useful in identifying a considerable number of isomeric benzenes, indans, and tetralins in a coal-derived jet fuel after electron ionization had proven unsuccessful.

The middle section of papers in this book deals with the application of novel mass spectrometric techniques to specific problems. For example, Johnson and Yost have used tandem mass spectrometry to study the MS/MS characteristics of steranes. EI MS/MS was shown to be more sensitive and selective than using PCI MS/MS. The sterane distribution in different oils could be determined by utilizing a parent scan of m/z 217. However, only the most abundant steranes could be confirmed by EI MS/MS daughter spectra. Some sample cleanup was suggested as the most viable approach to low-level detection of steranes and to prevent instrument contamination. Harriman et al. have compared two different methods for increasing the sensitivity of biomarker analysis beyond conventional capillary column GC/MS using nominal mass resolution. The two different methods involve (1) increasing the resolving power to 2000 to 3000 or (2) monitoring fragments formed in the first field free region of a double focussing mass spectrometer. The second method was shown to provide adequate sensitivity using whole oils without sample cleanup in addition to providing results which are easier to interpret and less susceptible to interferences than the first method. Beato and Yost investigated the electron ionization and ammonia chemicalionization behavior of metalloporphyrins. The purpose of this study was to correlate the degree of fragmentation with the metal involved in the complex. Their results were correlated with the stability index, $S_I = [100(PE)(V)]/R$, where PE is the Pauling electronegativity of the metal, *V* is the valence, and *R* the ionic radius. Metalloporphyrins with high stability indices gave intense doubly charged ions in EI and little fragmentation in CI. On the other hand, metalloporphyrins with low stability indices gave less abundant doubly charged ions in EI and fragmented to mono- and dipyrrolic fragments in CI. Gallegos reports the analysis of *n*-alkylbenzothiophenes by GC/MS. The close similarity between identified *n*-alkylbenzothiophenes and *n*-alkylbenzenes in an Arabian heavy Crude suggests the sulfur analog is formed from the latter. This reaction is postulated as occurring during maturation of the oil when *n*-alkylbenzenes come in contact with sulfur. Schmidt and Sprecher report the utilization of a low voltage high-resolution mass spectrometric direct insertion probe method for determining the heteroatomic composition, homologous series and carbon number distribution of coal-derived materials. Application of this method to a study investigating the removal of organosulfur compounds from coal extracts treated with molten caustic is described. Crawford et al. describe an on-line method to study the kinetics and chemistry of formation and evolution of ammonia during pyrolysis of oil shale. Isobutane chemical ionization was used because of its sensitivity and selectivity for ammonia overcoming possible interferences with water. Taghizadeh et al. investigated the combination of low voltage mass spectrometry with multivariate analysis for on-line process monitoring of coal-derived liquids. This method was applied to compositional changes which occur during hydrotreatment using fixed bed catalysts. As expected the major trend was the "hydrotreatment effect;" an increase in the percentage of hydroaromatics relative to aromatics in the product stream. Secondarily, with increased time and temperature, more higher molecular weight components, primarily alkyl substituted aromatics, were detected.

The last two papers contain a significant component of computer programming expertise to facilitate the mass spectrometry data/identification interface. Aczel et al. describe development of an expert system for interpretation of the GC/MS spectra of fossil fuel distillates. This system uses interpretation rules generally carried out by a skilled professional, such as the deconvolution of peaks, sets acceptability limits for commercial data system identifications and identifies components not identified by commercial data systems. The need for such a system stems from the need to identify complex hydrocarbon mixtures. Results indicate identifications are correct 90 to 95% of the time and result in a considerable reduction in skilled interpretation time. In the final paper Grigsby describes the processing of high resolution mass spectral data using Kendrick masses. The computer programs used for this method enable elemental formulas to be assigned to ions in high resolution mass spectra and displayed in a rectangular array to facilitate identification. Results are presented for a selected fraction containing one-ring aromatic neutrals from a Venezuelan heavy petroleum.

The papers presented in this book provide the reader with an overview of and some insight into the types of ionization, separation and computer techniques which will become routine in the near future. The editors would like to acknowledge the efforts of the authors as well as the ASTM personnel who have made publication of this book possible.

Karl V. Wood

Purdue University, West Lafayette, IN 47907; symposium chairman and editor.

Terrence R. Ashe

Esso Petroleum Canada, Sarnia, Ontario, Canada N7T 7MI; symposium chairman and editor.

Edward K. Chess,[1] Henry T. Kalinoski,[1] Bob W. Wright,[1] Harold R. Udseth,[1] and Richard D. Smith[1]

Applications of Supercritical Fluid Chromatography-Mass Spectrometry in the Analysis of Fossil Fuels

REFERENCE: Chess, E. K., Kalinoski, H. T., Wright, B. W., Udseth, H. R., and Smith, R. D., **"Applications of Supercritical Fluid Chromatography-Mass Spectrometry in the Analysis of Fossil Fuels,"** *Novel Techniques in Fossil Fuel Mass Spectrometry, ASTM STP 1019,* T. R. Ashe and K. V. Wood, Eds., American Society for Testing and Materials, Philadelphia, 1989, pp. 5–23.

ABSTRACT: Supercritical fluid chromatography-mass spectrometry (SFC-MS) and related techniques have been applied to the chemical characterization of a variety of fossil fuels, including diesel fuels, diesel fuel sediments, and petroleum- and coal-derived liquids and solids. Supercritical fluid chromatography can provide either high resolution or extremely rapid separation of complex fuel-derived mixtures. Insoluble sediments from marine diesel fuels can be extracted with polar supercritical fluids and the extracts directly introduced to a mass spectrometer for on-line characterization. Chemical class separations and mass spectrometric analysis of coal-derived liquids can be accomplished by utilizing high-performance liquid chromatography columns and a specially designed interface to the mass spectrometer. The development of polar supercritical fluids and chromatographic columns for use with the polar fluids and the development (in addition to refinement) of the SFC-MS interfaces and capillary flow restrictors have been responsible for the successful application of SFC-MS and related techniques to the analysis of fossil-fuel-related materials.

KEY WORDS: supercritical fluid chromatography, mass spectrometry, supercritical fluid chromatography-mass spectrometry, supercritical fluid extraction, direct fluid injection, fossil fuels, chemical class fractions, diesel fuel sediments, coal-derived liquids, characterization

In recent years, an active growth in the variety of applications of supercritical fluid chromatography (SFC) in the field of analytical chemistry has occurred. Because the physical properties of a supercritical fluid (for example, low viscosity, high diffusivity, reasonable solvating power, etc.) impart characteristics to the supercritical mobile phase intermediate to those of a liquid and a gas, SFC has allowed the high-resolution or high-speed separation of low-volatility or thermally labile components of some complex mixtures that were not separable by gas chromatography (GC) [*1–10*]. The flame ionization detector (FID) for SFC provides universal detection with supercritical mobile phases such as carbon dioxide (CO_2), but is precluded from use with polar-modified fluids, which may contain up to 20% by volume of a polar modifier, such as methanol, 2-propanol, or acetic acid. Conventional ultraviolet (UV) absorption detection suffers from sensitivity limitations for capillary SFC

[1]Senior research scientist, research scientist, senior research scientist, research scientist, and staff scientist, repectively, Chemical Sciences Department, Pacific Northwest Laboratory, Richland, WA 99352; Dr. Kalinoski is currently at Lever Brothers Research and Development, Edgewater, NJ 07020, Dr. Chess is currently at Baxter Healthcare Corporation, Round Lake, IL 60073.

and has further limitations both on the scope of application and the range of amenable fluids. Mass spectrometry (MS) has provided a sensitive detector for capillary and packed column SFC that can readily accommodate polar-modified fluids and provide a more information-rich spectral output than most other types of detectors. The fundamentals and practice of SFC-MS have been recently reviewed [*11*].

During the past five years, SFC has been applied to the analysis of fossil fuels through various approaches. Of the 100 citations found in a literature search of the Energy Research Abstracts data base (1981 through 1986) for the term "supercritical fluid chromatography," 76 citations involved research that employed a detector other than a mass spectrometer. For example, capillary column SFC-FID has been used in the analyses of petroleum- and coal-derived mixtures [*12–15*] and in the analysis of diesel fuels [*16*]. Hydrocarbon types found in petroleum liquids have been quantified by packed-column SFC-FID [*17*]. Polynuclear aromatic hydrocarbons (PAH) from extracts of diesel particulates [*18*] and carbon black (studied in our laboratory) have been identified by SFC-UV detection. Analytical-scale supercritical fluid extraction (SFE), combined with on-line capillary GC, has been applied to the analyses of the extractable materials from air particulate matter and coal tar [*19*]. The SFC introduction of samples to Fourier-transform infrared spectrometers has been used to study nonvolatile materials derived from coal [*20,21*]. Capillary SFC-FID has recently been utilized for simulated distillation of high-boiling petroleum fractions having atmospheric equivalent boiling points of less than 774°C (1426°F) [*22*]. This latter work by Schwartz et al. is a good example of how comparable information derived from GC analyses requiring temperatures >400°C (>752°F) can be obtained by SFC with linear pressure programming at a temperature of 100°C (212°F). The lower operating temperature had the added benefit of not thermally decomposing some components of the sample that were known to be affected at the higher GC column temperatures.

The application of SFC-MS to the analysis of fossil fuels has been demonstrated primarily in our laboratories at Pacific Northwest Laboratory. Our research programs have several focuses, all of which are necessary for the successful application of SFC-MS to the analysis of fossil fuels. These directions include: (1) the development of polar supercritical fluids and chromatographic columns for use with the polar fluids; (2) the development and refinement of the SFC-MS interface and capillary flow restrictors; and (3) the development of experimental procedures (such as SFC, direct fluid injection (DFI), and supercritical fluid extraction (SFE)) that can be applied to yield analytically useful information about fossil-fuel-related materials existing in a wide variety of matrices. This paper will primarily describe research conducted in our laboratories in these three research areas.

Experimental Procedure

Equipment—Details of the equipment used for capillary SFC-MS and SFE-MS have been described previously [*1,2,10,23–25*]. Computer-controlled syringe pumps were used to generate pressure-regulated flows of purified and distilled fluids. Temperature regulation of the chromatographic columns and extraction cells was provided by Hewlett-Packard gas chromatograph ovens. Valco C14W high-performance liquid chromatography (HPLC) injection valves with rotor volumes of 60 nL were used for sample introduction for capillary column SFC experiments. Capillary columns were prepared as previously described [*1–3*].

Microbore SFC separations were performed with a 1.0 by 100 mm microbore HPLC column packed with 5-μm silica particles that were coated with a C_{18}-bonded phase (ODS-Hypersil). To minimize dead volumes, the column was placed directly inside the 1.27-cm-outside diameter (OD) interface probe. A Rheodyne Model 7520 HPLC injection valve with a 0.2-μL injection volume was connected to the column with a length of deactivated 50-μm-inside diameter (ID) fused silica capillary tubing no longer than 20 cm.

Procedure—Specimens of fossil-fuel-related materials were prepared for SFC-MS analysis by dissolution of the solids in a minimum amount of a suitable solvent, such as methylene chloride. Extractions were conducted by placing microgram to milligram amounts of solids in a stainless steel extraction cell and by using various fluids as previously described [*26*]. Capacity ratio measurements using polar probe molecules and polar-modified CO_2 fluid systems were also conducted as previously described [*27*].

Results and Discussion

Development of Polar Fluids and Columns

One research focus has been on developing polar supercritical fluids and chromatographic columns that can be used for the characterization of higher-molecular-weight (above 250 dalton) and polar components of fossil fuels. Our research indicated that supercritical CO_2 is a relatively poor solvent for polynuclear aromatic hydrocarbons (PAH) and polar-substituted PAH, and applications are facilitated by (at a minimum) modification by the addition of a few percent of a polar solvent, such as 2-propanol, in order to solvate polar fossil fuels. This is because the solvating power (as the solubility parameter) of a supercritical fluid mobile phase (to a first approximation) is directly proportional to the density and is dependent on the chemical characteristics of the fluid. Intrinsically higher solvating powers and specific chemical selectivities can be obtained by selecting fluids with specific chemical properties for the analytes of interest (for example, presence of polar functional groups, pi bonds, etc.). Unfortunately, the critical temperatures for the more polar solvents that provide the capacity for specific or selective chemical interactions are prohibitively high (generally in excess of 250°C (482°F)) to be used as supercritical fluid mobile phases. However, mixed fluid systems, in which a polar modifier has been added to a fluid with a low critical temperature, are often characterized by increased solvating power (compared to the nonpolar fluid) and moderate operating temperatures. In general, nonpolar fluids have relatively low critical temperatures and the addition of up to 20 mole percent of a polar modifier will not increase the critical temperature of the resulting binary mixture beyond a workable range. The best choice of fluid modifiers is generally a solvent that is itself a good solvent for the analyte(s).

The effects of polar modifiers on the solvating power of nonpolar supercritical fluid mobile phases in SFC have been studied [*27*]. Capacity ratio (k') measurements derived from chromatographic experiments using polar probe molecules and polar-modified CO_2 fluid systems have shown significant decreases in retention of the probe molecules, indicating increased mobile phase solvating power. The three polar modifiers used in these studies all acted to increase the solvating power of the mobile phase, with the 2-propanol creating the largest decrease in retention, followed by acetonitrile and methanol. A 20 mole percent modifier concentration was sufficient to drastically change the solvating properties of the CO_2, and yet the critical temperatures were maintained below 120°C (248°F). Considering the magnitude of the retention decreases, however, modifier concentrations lower than 20 mole percent would be used in most separations.

Capillary SFC-MS chromatograms obtained with a quadrupole mass spectrometer of a coal tar mixture using a 2-propanol modified propane (10% volume/volume) fluid system are shown in Fig. 1. These chromatograms were obtained at different chromatographic operating temperatures and illustrate the importance of understanding the phase behavior of binary fluid mixtures. The separation obtained at 120°C (248°F) shows very poor sepa-

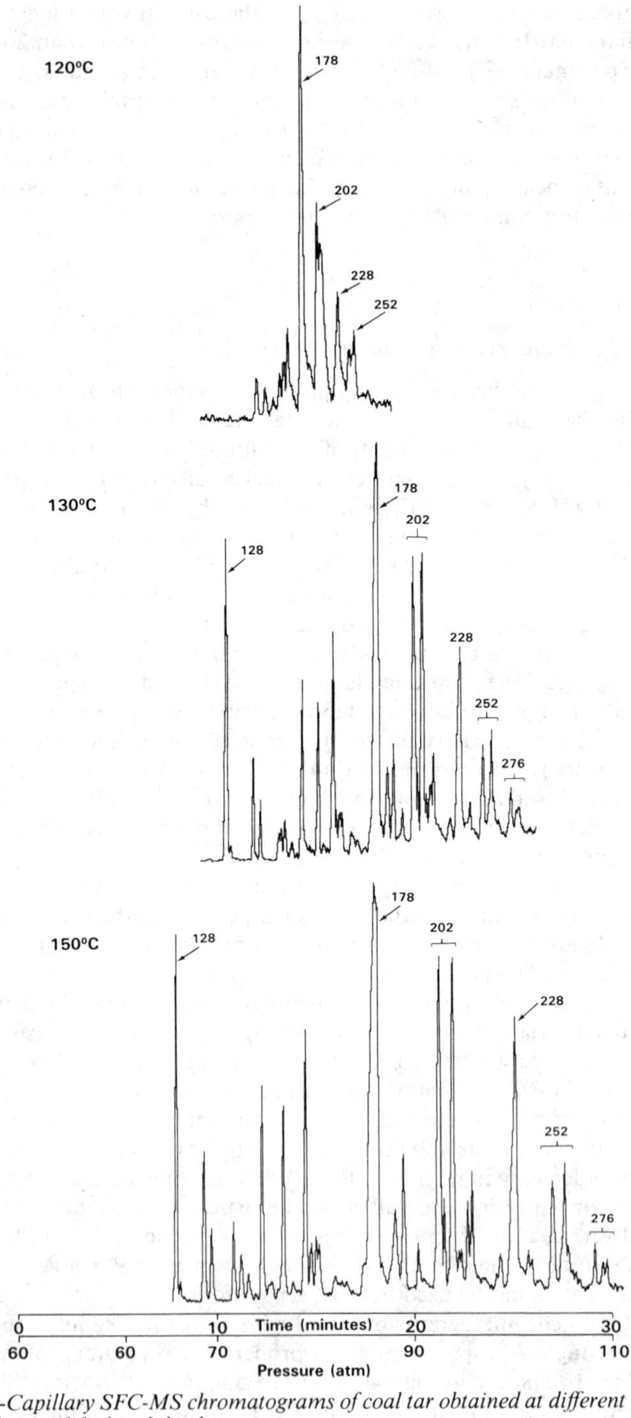

FIG. 1—*Capillary SFC-MS chromatograms of coal tar obtained at different temperatures using a polar-modified mobile phase.*

ration efficiency, with the components being eluted in an unresolved hump at lower pressures. Increasing the temperature by 10°C, to 130°C (266°F), dramatically improved the chromatographic performance, producing much narrower peaks and eluting the components over a wider pressure range. At the lower temperature, the fluid mixture was probably below its critical point and a subcritical liquid existed as the mobile phase. Increasing the temperature another 20 to 150°C (302°F) only slightly improved the chromatographic performance. This example not only illustrates the importance of understanding the critical phase behavior of a fluid mixture, but demonstrates the improved chromatographic performance of a supercritical mobile phase compared to a subcritical liquid mobile phase where pressure programming advantages are minimal.

The SFC-MS Interfaces and Flow Restrictors

Another primary focus of our research has been the development of the SFC-MS interface and flow restrictor technology. Successful interfaces have been developed for coupling capillary columns to both quadrupole [*23*] and magnetic sector mass spectrometers [*24*], and a unique high flow rate (HFR) interface for coupling packed microbore columns to a quadrupole instrument has also been developed [*28*]. These interfaces share some common design features: (1) the portion of the interface attached to the chromatograph is packaged as a heated probe that inserts into the vacuum chamber of the mass spectrometer; (2) the capillary flow restrictor is attached to the end of the capillary column or transfer line with a zero-dead-volume butt connector; and (3) the flow restrictor is contained at the end of the probe inside a stainless-steel capillary tube that can be resistively heated.

In addition, unique features have been found necessary for each different type of instrument. For example, the interface to the magnetic sector instrument requires that the probe tip and restrictor heater be electrically isolated from the chromatograph and interface probe shaft. The design permits the probe tip to be in contact with the ion source, which operates at >6000 V, and prevents high-voltage "arcing" that would preclude operation of the chemical ionization (CI) ion source. The HFR interface design is unique among the interfaces because it includes a region, into which the supercritical fluid expands, that is directly pumped independently from the remainder of the ion source chamber. Flow rates of up to 200 μL/min for compressed CO_2 have been obtained without any degradation of ion source performance with this interface design (compare with ~20 μL/min for conventional SFC-MS interfaces).

Direct fluid injection (DFI) is based on the rapid expansion of the supercritical mobile phase into a region where ionization can occur. Since proper operating conditions produce a gas following the expansion, both conventional electron ionization (EI) and CI have been used; generally, CI has offered the best detection limits in our laboratory. (When CO_2 alone is an appropriate fluid for SFC, CO_2 charge exchange can often generate EI-like spectra when operating in the CI mode.) The fluid expansion process through the flow restrictor is an important step of the DFI technique, one that defines both the capabilities and ultimate limitations. The flow restrictor performance, then, constitutes the most important component of the SFC-MS interface. A wide variety of restrictor designs have been studied, each with its own advantages and drawbacks [*29*]. These designs include point restrictors, restrictors drawn to various lengths and internal diameters, and tortuous path restrictors. There exists no one ideal restrictor for all applications of capillary SFC-MS; however, current interface designs can be used nearly routinely for many applications and are amenable to most restrictor configurations.

Applications of SFC-MS and SFE-MS to Fossil Fuels

A variety of fossil fuels, including diesel fuels, diesel fuel sediments, coal-derived liquids and solids, and porphyrins, have been analyzed by SFC-MS, SFE-MS, and related techniques. Many of the materials studied to date have been compatible with characterization by GC-MS, but some materials contained components that were too polar or too nonvolatile for GC-MS analysis. It is the extension of the high-efficiency separations available by GC to higher-molecular-weight and more-polar compounds that is a primary goal in the application of SFC-MS to the analysis of fossil fuels.

SFC-MS Analyses—The aromatic hydrocarbon chemical class fraction of coal tar was among the first fuel-related materials analyzed by SFC [*30*], and it has served as a standard test material during the development of SFC-MS and SFE-MS. Important among these developments was the use of short columns and steep pressure gradients to produce rapid, efficient separations of complex mixtures. For example, the chromatograms of the separation of the same coal tar sample using two different pressure gradients are shown in Fig. 2. The high-resolution separation (Fig. 2*a*) was performed utilizing supercritical CO_2 at 60°C (140°F) and a linear pressure ramp from 75 to 175 bar at 1 bar/min with a 15 m by 50 μm ID column. The faster separation (Fig. 2*b*) was obtained with a 1.75 m by 50 μm ID column with a pressure ramp of 50 bar/min starting at 75 bar. There is a noticeable loss of chromatographic separation in the faster separation, but the dramatic reduction in time necessary for the analysis may be more important in many analytical applications. The inherent selectively afforded by the mass spectrometer used as the detector can be used to overcome some of the loss of chromatographic resolution experienced in the rapid separation [*2*].

An example of the capability of high-resolution capillary column SFC-MS for the analysis of complex mixtures is shown in Fig. 3, the chromatogram of a separation of the polycyclic aromatic hydrocarbon class fraction of a marine diesel fuel [*31*]. This analysis was performed with CO_2 at 60°C (140°F) and a linear pressure ramp. The isobutane CI produced predominantly $[M + H]^+$ species, with smaller amounts of $[M + C_3H_7]^+$ and $[M + C_4H_9]^+$. The chromatographic peaks are annotated with the nominal molecular weights of the individual mixture components. The solvent peak at ~10 min is small because the molecular weight of the solvent was below the low mass end of the scan range.

DFI-MS and SFE-MS Analysis—Because a mass spectrometer can be used to characterize (fingerprint) mixtures and even identify mixture components in the absence of a chromatographic separation step, we have used supercritical fluids to transport materials directly to the ion source through a short transfer line in a technique we refer to as direct fluid injection (DFI). Chemical class fractions of marine diesel fuels were introduced to the ion source by DFI using CO_2 and ammonia as supercritical solvents. Mass spectra were acquired by means of isobutane or ammonia CI. Even though some of the class fractions examined contained materials too polar to be analyzed by SFC-MS, the mass spectra acquired by DFI-MS appeared to reflect of the range of compounds present, and allowed rapid, accurate determinations of average molecular weight to be made [*16*].

Polar sediments from marine diesel fuels were examined by SFE-MS, using ammonia as the extracting fluid, followed by on-line tandem mass spectrometric analysis using a triple-quadrupole mass spectrometer and collision-induced dissociation [*32*]. These insoluble sediments had formed upon aging of the fuels, and had proven fairly intractable toward conventional mass spectrometric analysis. Ammonia [T_c = 133°C (271°F), P_c = 113 bar] and methylamine [T_c = 157°C (315°F), P_c = 75 bar] proved to be very good solvents for these materials, resulting in mass spectra containing prominent molecular ions. Figure 4 depicts representative methane CI mass spectra of this sediment obtained during SFE-MS

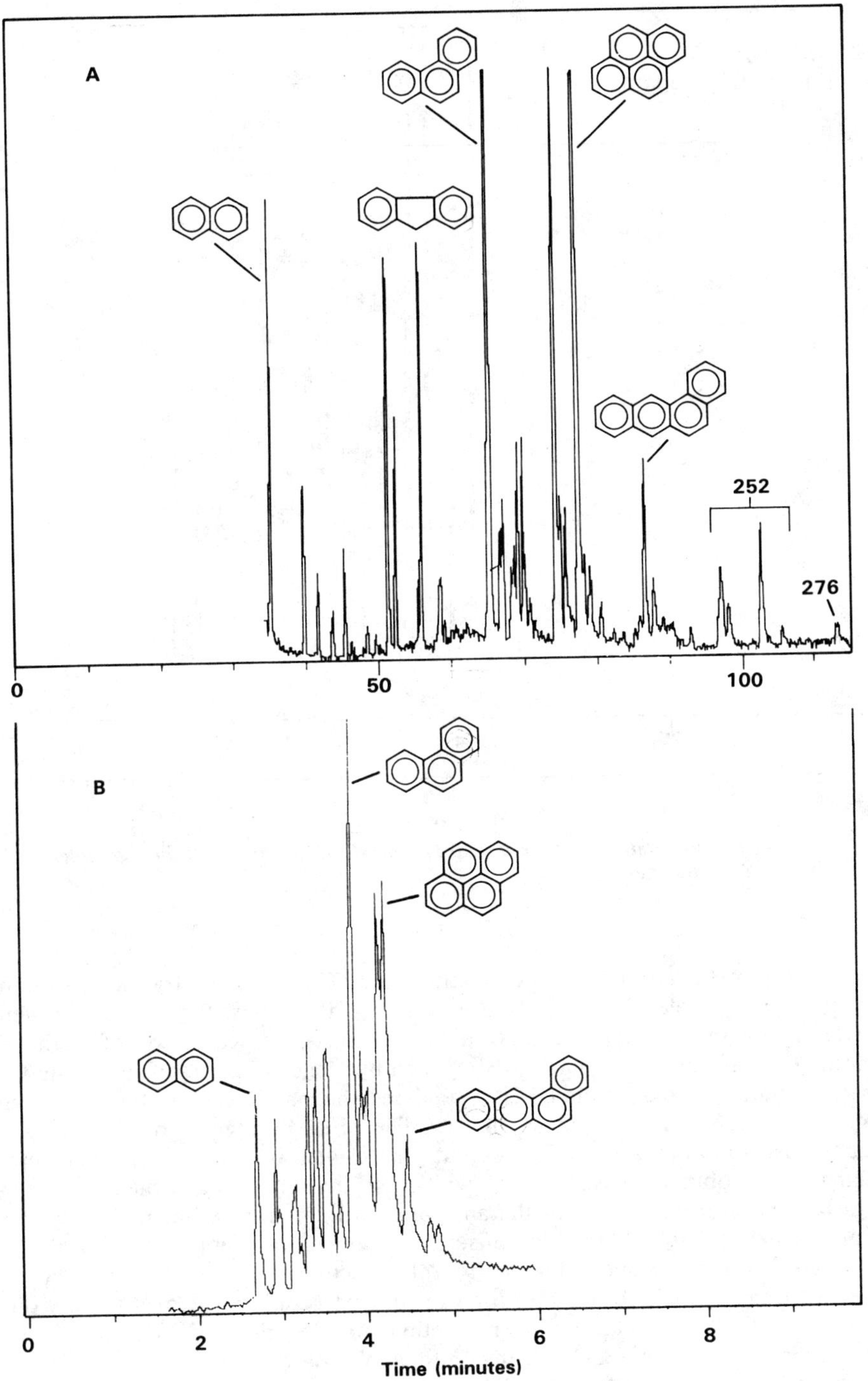

FIG. 2—*Capillary SFC-MS chromatograms of coal tar obtained with different pressure gradients.*

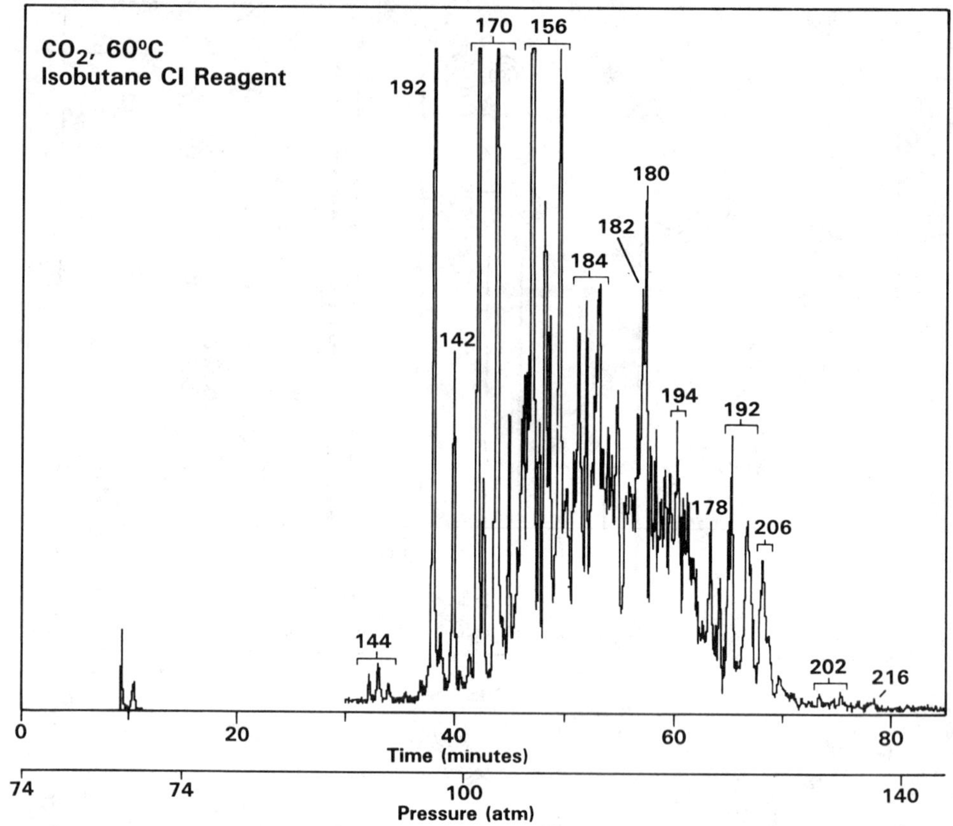

FIG. 3—*Total ion current chromatogram of an SFC-MS separation for a complex PAH fraction of a marine diesel fuel.*

analysis using methylamine as the extracting fluid. The mass spectra indicate sediment components with molecular weights in excess of 600 daltons. Because the presence of methylamine in the ion source did not appear to affect the acquisition of methane and ammonia CI mass spectra of standards injected by means of a methylamine fluid under the same conditions used for the DFI-MS analyses, the mass spectra in Fig. 4 are assumed to accurately represent methane CI spectra. The SFE-MS analysis of similar sediment materials, using ammonia as the CI reagent gas, resulted in mass spectra that were very similar to those obtained by extraction with methylamine. Tandem mass spectrometric analysis of the prominent ions in the ammonia extraction experiment showed that the sediment material that had formed in unstable diesel fuel was composed primarily of alkyl indoles, alkyl carbazoles, and polymeric pyrrole structures.

Acetic-acid-modified CO_2 was also used for the extraction of the diesel fuel sediments successfully extracted with ammonia and methylamine. Similar sediments had previously been examined by on-line SFE-MS using 2-propanol-modified CO_2, and the results of these extractions indicated that little material was being extracted and transported to the mass spectrometer by this fluid mixture. The acetic-acid-modified CO_2, however, extracted and transported material to the mass spectrometer that produced ions with molecular weights

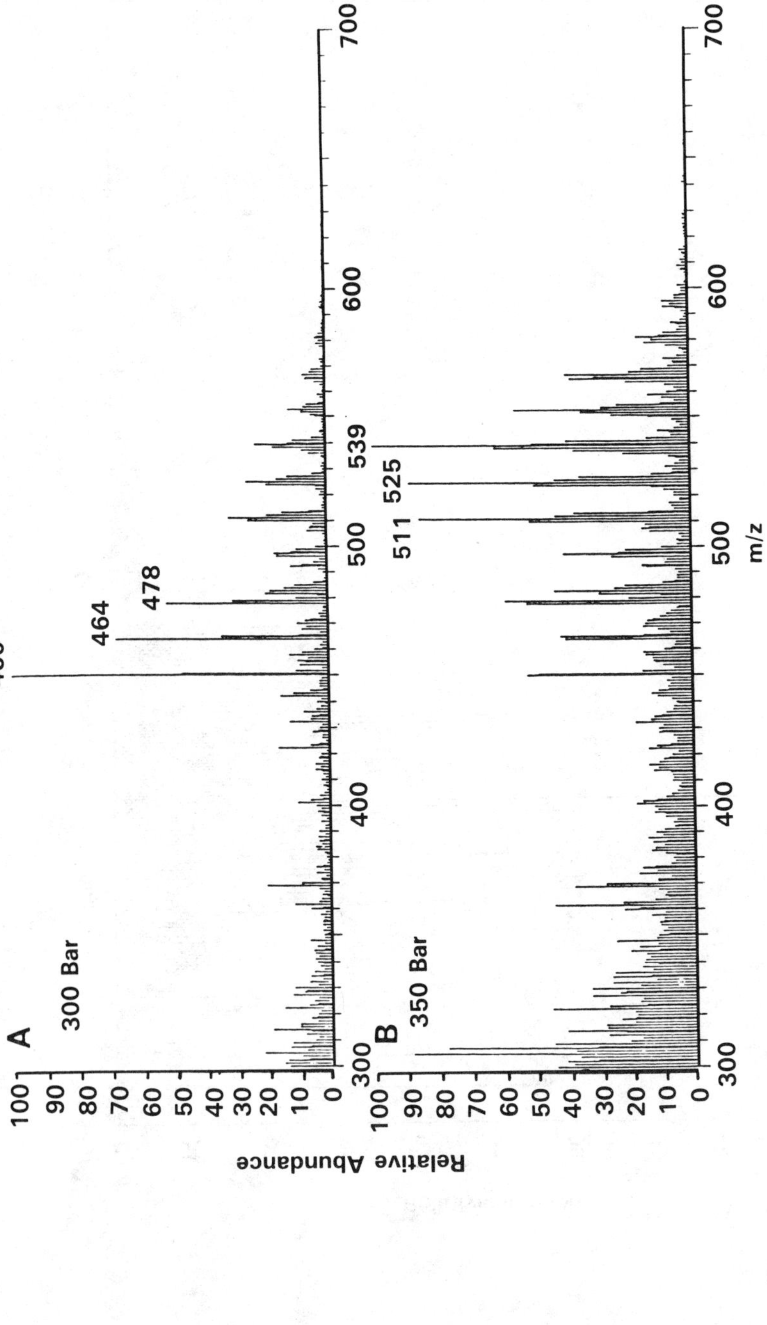

FIG. 4—*Methane CI mass spectra of a marine diesel fuel sediment acquired by SFE-MS analysis using methylamine at 165°C (329°F) and with fluid pressures of (A) 300 bar, and (B) 350 bar.*

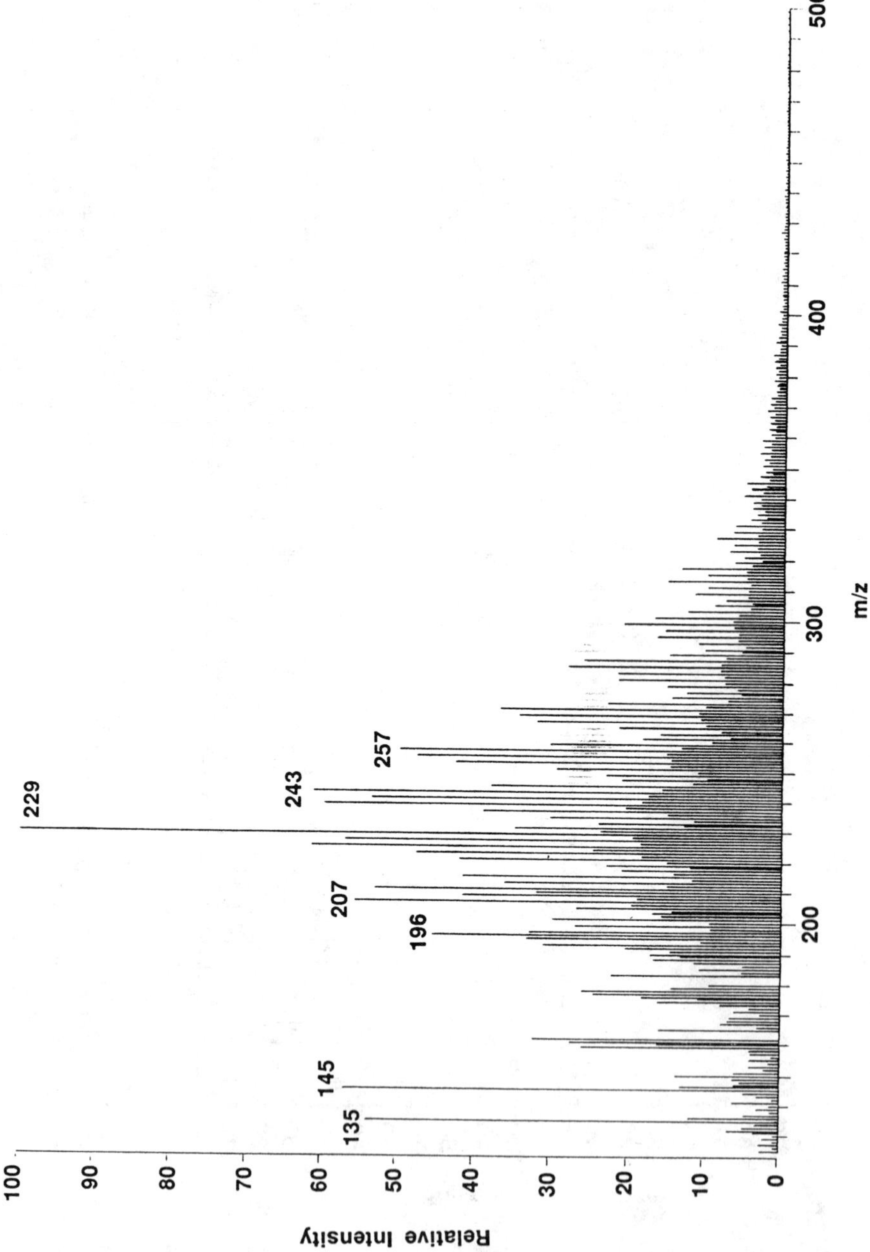

FIG. 5—*The CI mass spectra of a marine diesel fuel sediment acquired by SFE-MS analysis using a 5.6% (volume/volume) acetic acid/CO_2 fluid at 148°C (298°F) and with fluid pressure of 350 bar.*

in excess of 600 daltons. Figure 5 is an averaged CI mass spectrum of the acetic acid-modified CO_2 (5.6% acetic acid, by volume) extract of a diesel fuel sediment acquired at an extracting pressure of 350 bar. The masses at m/z 163 to 165 and m/z 181 to 183 were not acquired because of the large signals originating from the cluster ions $(CH_3COOH)_3H^+$ (m/z 181) and $[(CH_3COOH)_3H\text{-}H_2O]^+$ associated with the large abundance of acetic acid. This sediment sample extract contained materials with molecular weights as large as 400 daltons. The difference in molecular weight distributions between these results and those from extractions performed with basic solvents may reflect the distributions of different chemical classes of materials extracted from the sediment matrices by the two classes of extracting fluids.

Microbore SFC-MS—Even though the highest-resolution SFC is afforded by employing capillary columns, the analysis of complex mixtures of high-molecular-weight, relatively nonvolatile analytes by capillary column SFC is limited by a number of factors. Currently, nearly all capillary SFC employs split-injection techniques in which only a small portion of the sample is introduced to the chromatographic column and detector. Split-injection techniques allow the high efficiency of capillary SFC to be fully exploited, but limit reproducibility. These techniques are also constrained to small sample sizes because of the sample volume, the limited solubility of analytes in the sample solvents, and the potential for overloading thin-film capillary columns. The use of capillary flow restrictors in SFC is, in turn, limited by solute precipitation of nonvolatile analytes, particularly at the low flow rates typically encountered with capillary SFC. These limitations are not present for SFC conducted with conventional and microbore HPLC columns.

The HFR interface was developed to allow operation at high fluid flow rates, permitting the use of microbore HPLC columns for SFC and enhancing sensitivity for relatively nonvolatile compounds [*30*]. Compared with other capillary direct-injection interfaces, the HFR interface has the advantages of allowing SFC operation at high flow rates, providing efficient analyte transfer to the CI region, allowing unrestricted CI operation at optimum pressure, and maintaining normal CI reagent gas flow rates while requiring only a relatively minor increase in complexity.

The ability to operate at high fluid flow rates permits the use of packed microbore HPLC columns for SFC-MS. The disadvantages of packed columns (compared with capillary columns) include the lower chromatographic resolution and the greater retention of higher-molecular-weight and highly polar analytes. The advantages of this type of column accrue from its higher effective phase ratio compared with 50-μm-ID, fused silica capillary columns (greater amount of stationary phase per unit area), resulting in a greatly increased sample size. The larger sample can be injected on-column as a splitless injection, which enhances sensitivity and dynamic range for a given sample concentration. A large effective dynamic range is essential for the complete analysis of extremely complex mixtures associated with fuels and coal-derived liquids. The selectivity of a separation can be enhanced by the wide range of commercially available stationary phases found in HPLC columns.

The application of pressure programming over a limited range with packed-column SFC and a polar-modified fluid (a technique widely practiced with capillary columns) allows full utilization of the selectivity of the chromatographic stationary phase and the variable solvating power of the fluid, a key advantage of supercritical fluids. Short (100-mm) microbore (1- or 2-mm-ID) packed columns are required to limit the pressure drop across the column because SFC requires uniform pressures to maintain uniform solvating powers. All extra column volumes (or dead volumes) need to be minimized to eliminate band broadening and the loss of chromatographic efficiency. A mass spectrometer is an ideal detector, as it is virtually unaffected by the polar modifier and is selective and sensitive for detecting a wide variety of samples. Furthermore, mass spectrometry adds an extra dimension of selectivity with the ability to identify components not fully resolved chromatographically on the short packed columns.

The nitrogen-containing polycyclic aromatic hydrocarbon (NPAC) chemical class fractions [*33*] of 371 to 399°C (700 to 750°F) and 399 to 427°C (750 to 800°F) boiling point cuts of an SRC-II (Solvent Refined Coal-II) heavy distillate product were analyzed using microbore SFC-MS [*34*]. The NPAC chemical class fraction is composed primarily of nitrogen-containing polycyclic aromatic hydrocarbons, in which the nitrogen is present as primary, secondary, and tertiary amines, including the indole- and carbazole-NH and aza-N moieties. The mass spectra shown in Figs. 6*a* and *b* are the summations (averages) of all the mass spectra acquired during the SFC-MS analyses of these two boiling-point fractions, and represent an integrated view of all the material entering the mass spectrometer ion source from the column that was ionized under the conditions of the experiment. The chemical ionization conditions were such that protonated molecular species were formed with a minimum of fragmentation.

The compounds in the spectrum in Fig. 6*a* (371 to 399°C (700 to 750°F) fraction) with protonated molecular ions at m/z 182, 196, 210, 224, 238, 252, etc., can best be represented as members of the homologous series of alkylated carbazoles. The base peak (that is, the most abundant peak) at m/z 196 is attributed primarily to the protonated molecular ion of C_2-carbazoles. The homologous series of ions at m/z 194, 208, 222, 236, and 250 represent a series of alkylated phenanthridines, acridines or benzoquinolines. These two homologous series represent the most abundant species in this boiling-point chemical class fraction. Under the conditions of the experiment, 75°C (167°F) and 360 bar (isobaric), compounds with molecular weights up to approximately 300 daltons eluted from the C_{18} column. The carbazole and phenanthridine series are also found to be important species in the mass spectrum of the 399 to 427°C (750 to 800°F) boiling point cut (Fig. 6*b*). Because the average boiling point is higher, the most abundant alkylated carbazole is observed at m/z 224, a shift of two methylene units higher than the most abundant carbazole homologue found in the 371 to 399°C (700 to 750°F) fraction. The alkylphenanthridine series is now the most abundant homologous series in the spectrum. A third homologous series, at m/z 204, 218, 232, 246, and 260, is also important in this spectrum; it is probably composed of a series of alkylated azapyrenes and azafluoranthenes.

The spectra shown in Fig. 6 are representative of the type of information that can be obtained by direct-probe inlet mass spectrometry when the ionization conditions are chosen such that only molecular weight information is obtained. The use of a microbore chromatographic column enables additional information about a complex mixture to be obtained because of the chromatographic separation of components that results from the chemical and physical interactions of the components with the stationary phase of the column. This concept can be demonstrated by examining the individual mass spectra (or, in this case, scan-averaged mass spectra that are the averages of data acquired during a 0.5-min interval) acquired during the microbore SFC-MS analyses of the two SRC-II boiling-point cut chemical class fractions previously discussed. Figure 7 displays four averaged mass spectra that represent the components that eluted during the analyses of the 371 to 399°C (700 to 750°F) SRC-II NPAC fraction during the time periods of 7.0 to 7.5 min (Fig. 7*a*), 7.5 to 8.0 min (Fig. 7*b*), 8.8 to 9.4 min (Fig. 7*c*), and 9.5 to 10.0 min (Fig. 7*d*). Figure 8 displays five averaged mass spectra that represent the components that eluted during the analyses of the 399 to 427°C (750 to 800°F) SRC-II NPAC fraction during the time periods of 7.6 to 8.2 min (Fig. 8*a*), 9.4 to 10.0 min (Fig. 8*b*), 10.6 to 11.2 min (Fig. 8*c*), 11.8 to 12.4 min (Fig. 8*d*), and 13.0 to 13.6 min (Fig. 8*e*).

Whereas the total ion chromatograms obtained for these analyses show a relatively featureless, rapid increase in intensity to a broad peak and a gradual decrease back to baseline,

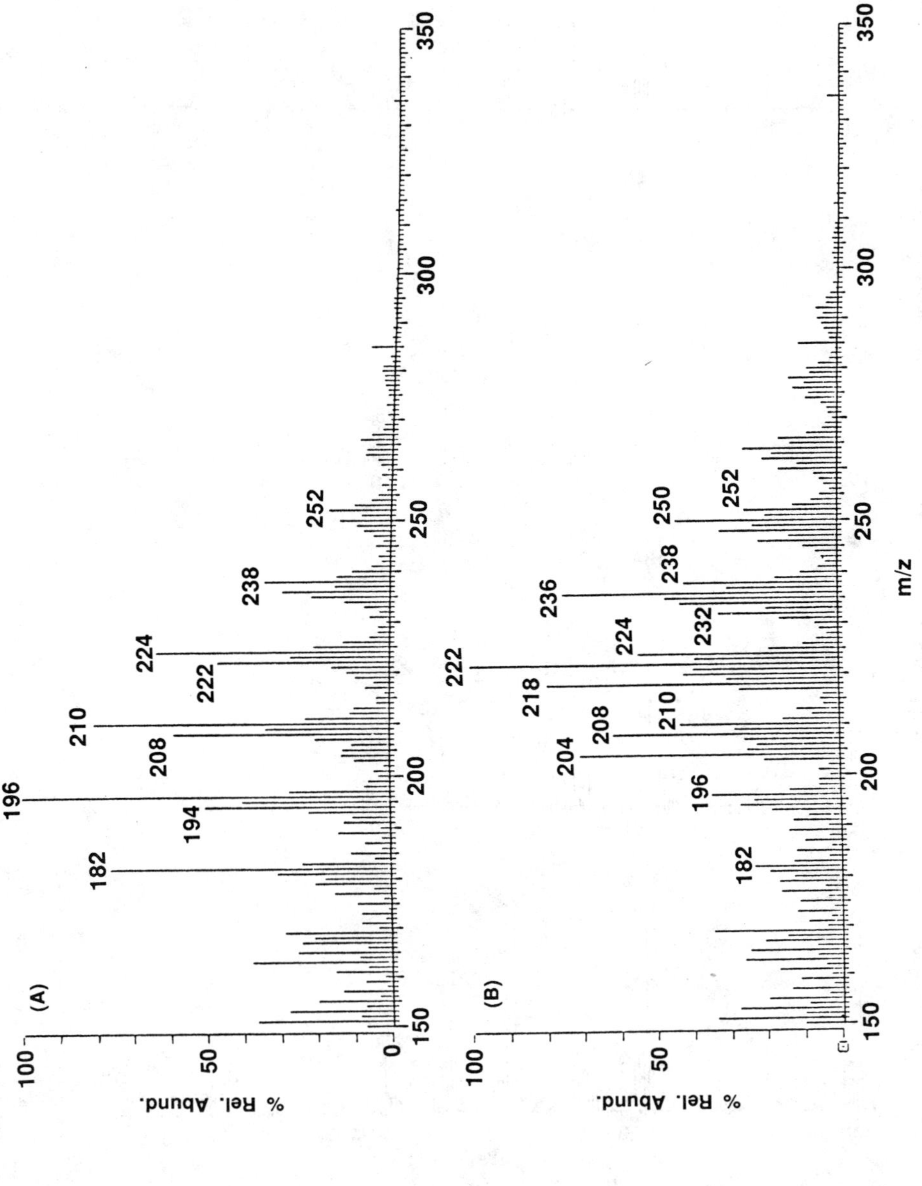

FIG. 6—*Scan-averaged mass spectra from the microbore SFC-MS analysis of NPAC chemical class fractions from the* (A) *371 to 399°C (700 to 750°F) and* (B) *399 to 427°C (750 to 800°F) boiling-point fractions of SRC-II heavy distillate product.*

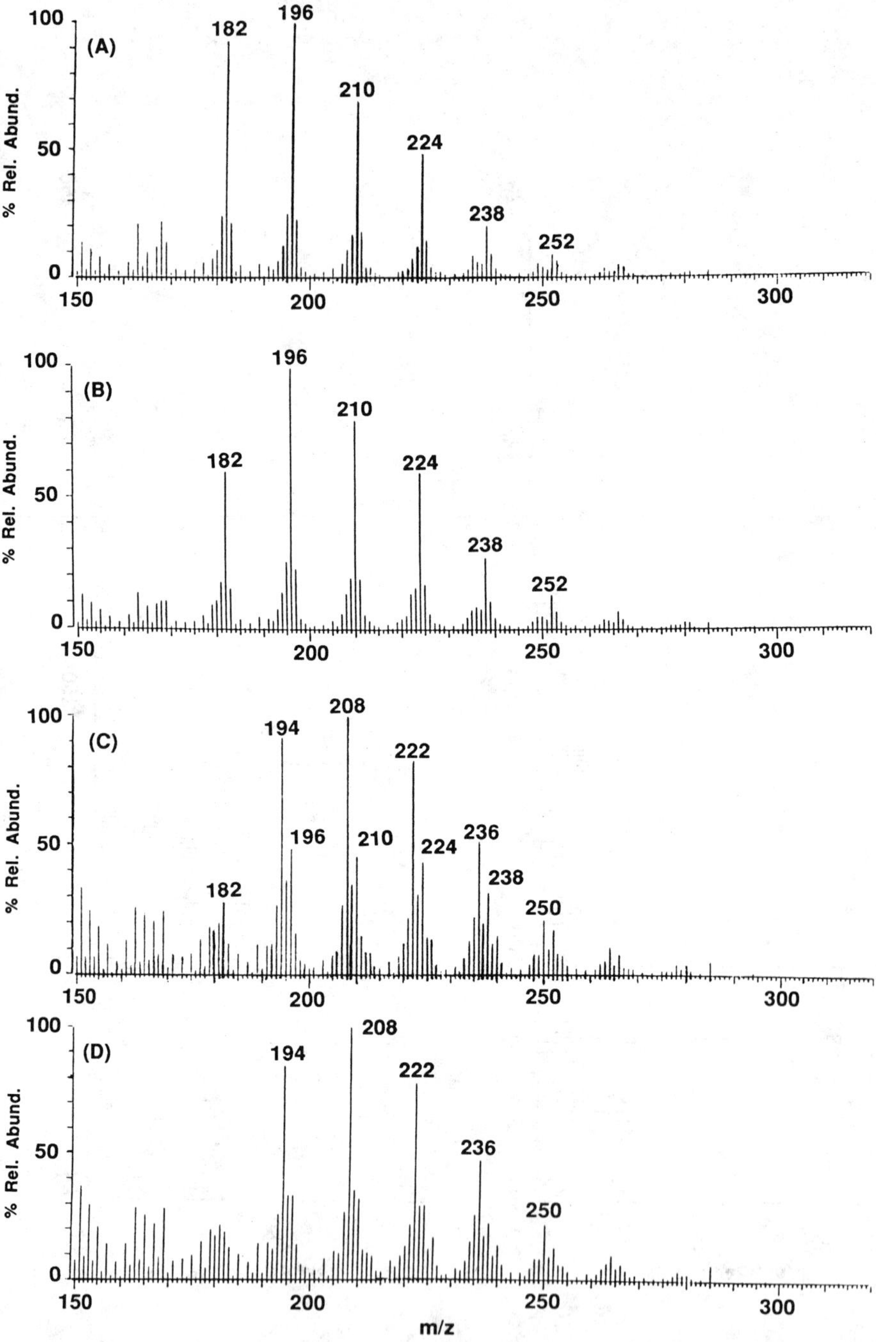

FIG. 7—*Mass spectra from the microbore SFC-MS analysis of the NPAC chemical class fraction, 371 to 399°C (700 to 750°F) boiling-point fraction, SRC-II heavy distillate product averaged over different acquisition time periods.*

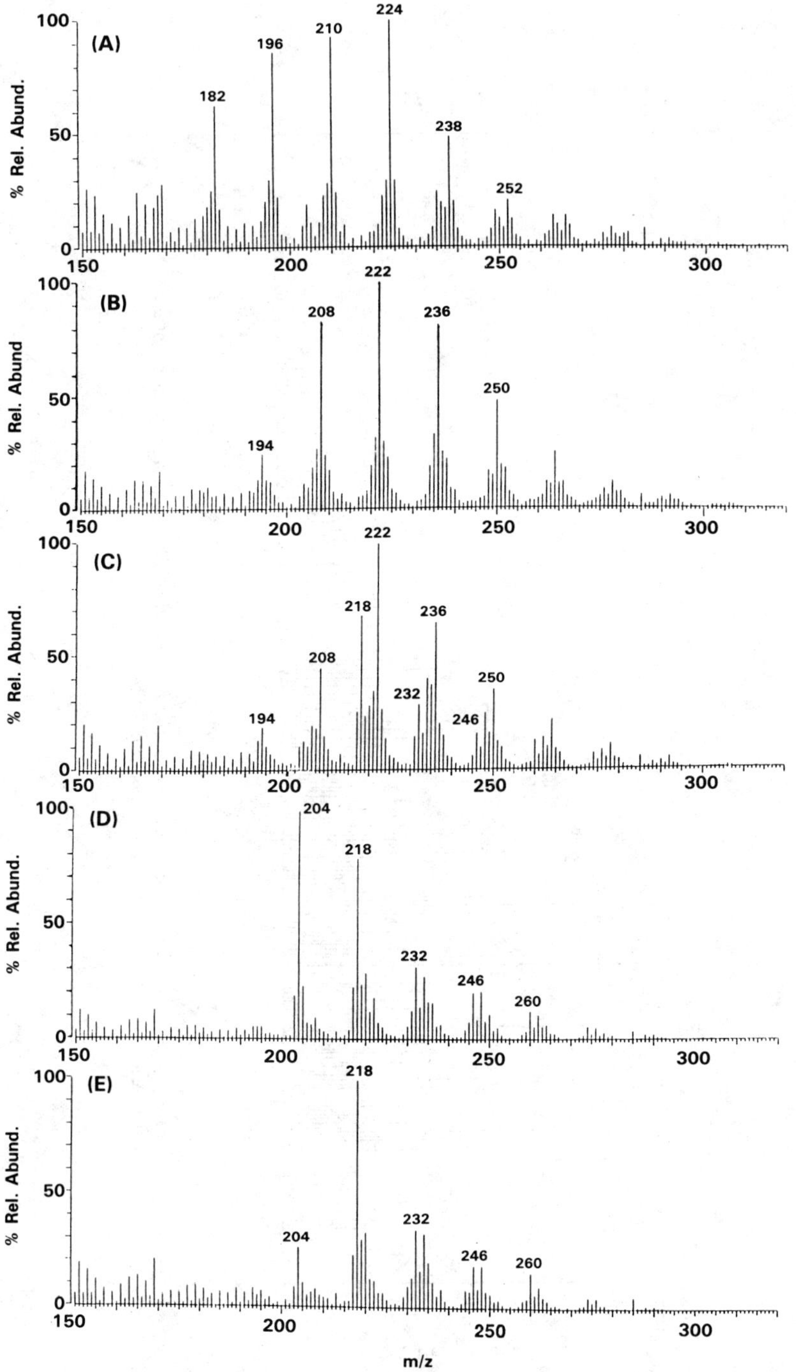

FIG. 8—*Mass spectra from the microbore SFC-MS analysis of the NPAC chemical class fraction, 399 to 427°C (750 to 800°F) boiling-point fraction, SRC-II heavy distillate product averaged over different acquisition time periods.*

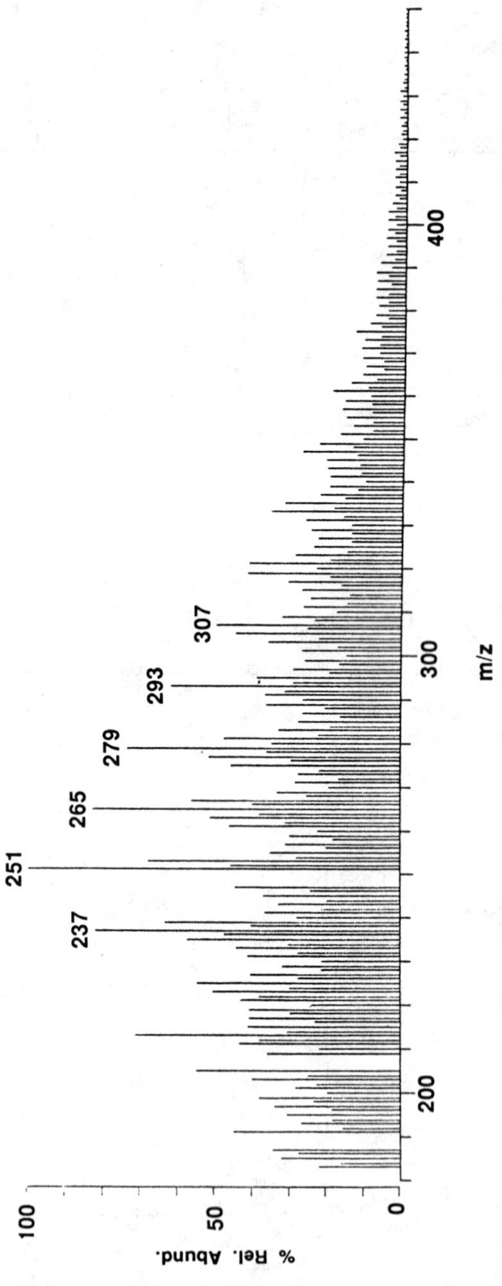

FIG. 9—*Scan-averaged mass spectrum of the hydroxy-PAC chemical class fraction of a hydrotreater unit feed from a coal liquefaction process.*

the individual spectra obtained during the elution show the changes in the chemical nature of the analytes being eluted. For example, the spectra shown in Fig. 7 show the carbazole homologous series (*m/z* 168, 182, 196, 210, etc.) eluting early in the analysis (Figs. 7*a* and *b*) and the quinoline homologous series (*m/z* 194, 208, 222, etc.) eluting later (Figs. 7*c* and *d*). A similar series of mass spectra, obtained for the 399 to 427°C (750 to 800°F) boiling-point cut, show the carbazole homologous series eluting first about 8 min (Fig. 8*a*), followed by the quinoline homologous series at about 9.5 to 10.0 min (Figs. 8*b* and *c*), followed by the azafluoranthene/azapyrene homologous series (*m/z* 204, 218, 232, etc.) and the benzocarbazole homologous series (*m/z* 218, 232, 246, etc.) eluting at about 12 min (Figs. 8*d* and *e*). This last series was not observed in the boiling-point cut of lower temperature. These data indicate that some degree of chemical class separation is occurring during the separation of the mixtures on the microbore packed column. However, the degree to which this mechanism of fractionation is responsible for the separations, relative to separations due simply to differences in volatility, has not been established.

The microbore SFC-MS technique has also been used to analyze a mixture containing polar components. Figure 9 shows the averaged CI mass spectrum of the hydroxy-containing polynuclear aromatic compounds (hydroxy-PAC) chemical class fraction from a coal liquefaction hydrotreater unit feed stream material. The analysis was conducted at 400 bar pressure and at 75°C (167°F) with the C_{18} column using 2% 2-propanol in CO_2. Contributions to the composite spectrum arising from sample components were present in excess of 400 daltons. This spectrum favorably compares with the mass spectrum of the same sample acquired by low-voltage electron impact, probe inlet mass spectrometry [*35*]. The small gaps observed in the mass spectrum in Fig. 9 (around *m/z* 194, 208, and 248) result because of the presence of intense background ions arising from solvent clusters and column bleed that prevented data from being acquired in those mass regions.

Conclusions

Supercritical fluid chromatography-mass spectrometry has the potential for making major contributions to fossil fuel analysis. For the potential to be realized, more knowledge about the ionization characteristics of unconventional CI reagents, such as polar-modified fluids, is needed. In addition, the phenomenon of solute nucleation in the SFC-MS interface region for nonvolatile compounds during expansion [*11*] requires more study in order to increase sensitivities. With this knowledge will come better designs for the interface itself. What this area of research needs most of all, however, are more active participants who can apply MS detection to some of the analyses already being performed by SFC, which will lead to more rapid progress in the area. The commercial availability of SFC instrumentation appears to be stimulating such participation.

Acknowledgment

Pacific Northwest Laboratory is operated for the U.S. Department of Energy by Battelle Memorial Institute. This work has been supported by the U.S. Department of Energy, Offices of Health and Environmental Research and Fossil Energy, and the U.S. Naval Research Laboratory under Contract DE-AC06-76RLO 1830.

References

[*1*] Wright, B. W., Kalinoski, H. T., and Smith, R. D., *Analytical Chemistry,* Vol. 57, 1985, pp. 2823–2829.

[2] Smith, R. D., Kalinoski, H. T., Udseth, H. R., and Wright, B. W., *Analytical Chemistry,* Vol. 56, 1984, pp. 2476–2480.
[3] Smith, R. D., Udseth, H. R., and Wright, B. W., *Journal of Chromatographic Science,* Vol. 23, 1985, pp. 192–199.
[4] Smith, R. D., Udseth, H. R., and Wright, B. W. in *Supercritical Fluid Technology,* J. M. L. Penninger, M. Radosz, M. A. McHugh, and V. J. Krukonis, Eds., Elsevier, Amsterdam, 1985, pp. 191–223.
[5] Chester, T. L., Innis, D. P., and Owens, G. D., *Analytical Chemistry,* Vol. 57, 1985, pp. 2343–2247.
[6] Novotny, M. and Springston, S. R., *Journal of Chromatography,* Vol. 279, 1983, pp. 417–422.
[7] Chester, T. L., *Journal of Chromatography,* Vol. 299, 1984, pp. 424–431.
[8] Wright, B. W., Udseth, H. R., Smith, R. D., and Hazlett, R. N., *Journal of Chromatography,* Vol. 314, 1984, pp. 253–262.
[9] Smith, R. D., Fjeldsted, J. C., and Lee, M. L., *Journal of Chromatography,* Vol. 247, 1982, pp. 231–243.
[10] Smith, R. D. and Udseth, H. R., *Analytical Chemistry,* Vol. 55, 1983, pp. 2262–2272.
[11] Smith, R. D., Kalinoski, H. T., and Udseth, H. R., *Mass Spectrometry Review,* Vol. 6, No. 4, 1987, pp. 445–496.
[12] Lee, M. L., "Characterization of Coal Macerals by High Resolution Supercritical Fluid Chromatography," Annual Report. PB-87-102158/XAB, National Technical Information Service, Springfield, VA, 1986.
[13] Lee, M. L., "Separation of Large Coal Molecules Using High-Resolution Supercritical Fluid Chromatography," Final Report. PB-86-132842/XAB, National Technical Information Service, Springfield, VA, 1985.
[14] Campbell, R. M. and Lee, M. L., *Analytical Chemistry,* Vol. 58, 1986, pp. 2247–2251.
[15] Wright, B. W. and Smith, R. D., *Chromatographia,* Vol. 18, No. 10, 1984, pp. 542–545.
[16] Wright, B. W., Kalinoski, H. T., Kopriva, A. J., Udseth, H. R., and Smith, R. D., "Characterization of Navy Fuels Using Supercritical Fluid Analytical Methodologies," Final Report, Defense Technical Information Service, Alexandria, VA, 1986.
[17] Norris, T. A. and Rawdon, M. G., *Analytical Chemistry,* Vol. 56, 1984, pp. 1767–1769.
[18] Jinno, K., Hoshino, T., Hondo, R., Saito, M., and Senda, M., *Analytical Chemistry,* Vol. 58, 1986, pp. 2696–2699.
[19] Wright, B. W., Frye, S. R., McMinn, D. G., and Smith, R. D., *Analytical Chemistry,* Vol. 59, 1987, pp. 640–644.
[20] Taylor, L. T., "Identification of Nonvolatile Coal Derived Products via Chromatography Coupled with On-Line FTIR Detection," Quarterly Technical Report, DOE/PC/70799-T8, National Technical Information Service, Springfield, VA, 1987.
[21] Pariente, G. L. and Griffiths, P. R., "Supercritical Fluid Extraction of Coal: Apparatus and Characterization Techniques" in *Proceedings* Pittsburgh Conference and Exposition on Analytical Chemistry and Applied Spectroscopy, Pittsburgh, PA, 1986.
[22] Schwartz, H. E., Brownlee, R. G., Boduszynski, M. M., and Su, F., *Analytical Chemistry,* Vol. 59, 1987, pp. 1393–1401.
[23] Wright, B. W., Kalinoski, H. T., Udseth, H. R., and Smith, R. D., *Journal of High Resolution Chromatography and Chromatography Communications,* Vol. 9, 1986, pp. 145–153.
[24] Kalinoski, H. T., Udseth, H. R., Chess, E. K., and Smith, R. D., *Journal of Chromatography,* Vol. 394, 1987, pp. 3–14.
[25] Kalinoski, H. R., Udseth, H. R., Wright, B. W., and Smith, R. D., *Analytical Chemistry,* Vol. 58, 1986, pp. 2421–2425.
[26] Smith, R. D., and Udseth, H. R., *Separation Science and Technology,* Vol. 18, No. 3, 1983, pp. 245–252.
[27] Chess, E. K., Kalinoski, H. R., Wright, B. W., and Smith, R. D., "Development and Evaluation of Supercritical Fluid Chromatography-Mass Spectrometry for Polar and High-Molecular-Weight Coal Components, Annual Report," PNL-6140, Pacific Northwest Laboratory, National Technical Information Service, Springfield, VA, 1987.
[28] Smith, R. D. and Udseth, H. R., *Analytical Chemistry,* Vol. 59, No. 1, 1987, pp. 13–22.
[29] Smith, R. D., Fulton, J. L., Petersen, R. C., Kopriva, A. J., and Wright, B. W., *Analytical Chemistry,* Vol. 58, 1986, pp. 2057–2064.
[30] Fjeldsted, J. C., Kong, R. C., and Lee, M. L., *Journal of Chromatography,* Vol. 279, 1983, pp. 449–455.
[31] Smith, R. D., Yonker, C. R., Kalinoski, H. T., Chess, E. K., Udseth, H. R., Frye, S. L., and Wright, B. W. in *Health and Environmental Research on Complex Organic Mixtures,* R. H. Gray,

E. K. Chess, P. J. Mellinger, R. G. Riley, and D. L. Springer, Eds., 24th Hanford Life Sciences Symposium, 21–24 Oct. 1985, Richland, WA, 1988, pp. 11–26.

[*32*] Wright, B. W., Kalinoski, H. T., Udseth, H. R., and Smith, R. D. in *Proceedings,* 2nd International Conference on Long-Term Storage Stabilities of Liquid Fuels, Vol. 2, L. L. Stavinoha, Ed., Southwest Research Institute, San Antonio, TX, 1986, pp. 526–539.

[*33*] Later, D. W., Lee, M. L., Bartle, K. D., Kong, R. C., and Vassilaros, D. L., *Analytical Chemistry,* Vol. 53, 1981, pp. 1612–1620.

[*34*] Chess, E. K., Kalinoski, H. R., and Smith, R. D., "Development and Evaluation of Supercritical Fluid Chromatography-Mass Spectrometry for Polar and High-Molecular-Weight Coal Components, Annual Report," PNL-6168, Pacific Northwest Laboratory, National Technical Information Service, Springfield, VA, 1988.

[*35*] Wright, C. W., "Reconfigured, Close-Coupled Reconfigured, and Wyodak Coal Integrated Two-Stage Coal Liquefaction Process Materials from the Wilsonville Facility: Chemical and Toxicological Evaluation," PNL-6159, Pacific Northwest Laboratory, National Technical Information Service, Springfield, VA, 1987.

Michael J. Gehron[1] *and Richard A. Yost*[1]

Hydrocarbon-Type Analysis of Jet Fuel with Gas Chromatography/Mass Spectrometry

REFERENCE: Gehron, M. J. and Yost, R. A., **"Hydrocarbon-Type Analysis of Jet Fuel with Gas Chromatography/Mass Spectrometry,"** *Novel Techniques in Fossil Fuel Mass Spectrometry, ASTM STP 1019,* T. R. Ashe and K. V. Wood, Eds., American Society for Testing and Materials, Philadelphia, 1989, pp. 24–37.

ABSTRACT: A gas chromatography/mass spectrometry hydrocarbon-type analysis (GC/MS/HTA) method has been developed for the analysis of jet fuels. The method employs a short-column (3 m by 0.25 mm inside diameter by 1.0 μm thick film) DB-5 capillary for sample introduction into the mass spectrometer. The calculational aspects of the technique are based on the current ASTM Test Method for Hydrocarbon Types in Low Olefinic Gasoline by Mass Spectrometry (D 2789-86) inverse calibration matrices. An evaluation of component discrimination with GC introduction was performed as a function of injection port temperatures and injection methods. The analytical figures of merit such as sensitivity and limit of detection were determined. Comparisons between the GC/MS/HTA and ASTM D 2789-86 results of several test fuels were made. Additionally, several synthetically prepared fuels were analyzed by the GC/MS/HTA method in order to evaluate the accuracy of the method.

KEY WORDS: gas chromatography/mass spectrometry, mass spectrometry, jet fuels, fuels, hydrocarbon-type analysis, discrimination, analysis of fuels, petroleum, fossil fuels

The need to understand the chemical composition of aviation fuels has never been more important than in the last decade. This need stems from two important aspects: the recent world-wide shortages of petroleum crude reserves that have resulted in a transition to alternative crude sources; and the desire to design and operate high performance jet engines with the knowledge of fuel composition as a controlling criterion. This has led to a resurgence in the development of new analytical techniques for petroleum analyses and more specifically aviation fuel analyses. In this laboratory, investigations are presently underway for the application of gas chromatography/mass spectrometry (GC/MS) for the analysis of aviation fuels (in particular, turbine jet fuels). These fuels represent a specific boiling point fraction or range of components of crude oil.

Type analysis in general does not provide information on specific individual components in a mixture, but rather supplies information on the relative amounts of certain classes of compounds. In many applications, the quantitation of specific compounds is not required, and indeed may provide much more information than is actually desired. This is particularly true in cases where the analyst is only interested in the overall macro-composition of a very complex mixture. The type analysis of hydrocarbon mixtures, such as gasoline and heavier oils, represents such an application. In 1951, Ralph Brown of the

[1]Research assistant and associate professor, respectively, Chemistry Department, University of Florida, Gainesville, FL 32611; Dr. Gehron is currently at Pratt and Whitney, West Palm Beach, FL 33410-9600.

Atlantic Richfield Corporation developed the first mass spectrometric technique for hydrocarbon-type analysis (HTA) [*1*]. Since the Brown method, there have been numerous improvements on his method due to improved calibration schemes [*2,3*] and improved instrumentation [*4,5*]. In 1962, standard HTA methods for gasoline and diesel fuels were adopted into the American Society of Testing and Materials (ASTM) standards. There have been several revisions of these methods over the years, but the basis of the methods remains the same. The ASTM Test Method for Hydrocarbon Types in Low Olefinic Gasoline by Mass Spectrometry (D 2789-86) and ASTM Test Method for Hydrocarbon Types in Middle Distillates by Mass Spectrometry (D 2425 83) perform the analysis on magnetic sector instruments that are no longer in commercial production. Because of this limitation, there is a much greater demand for samples to be analyzed than there are laboratories equipped to perform them. Additionally, the means of sample introduction into the mass spectrometer, which is via a heated glass inlet volume or microburette, is a time-consuming procedure and may facilitate component discrimination [*6*]. The technique also relies on manual measurements of peak height intensity that eliminates the ability to computerize the calculations. Consideration of these limitations illustrates the need for a modern HTA method performed on a widely available modern instrument. With the development of a GC/MS/HTA, several advantages would be realized. First, convenience in instrumentation; since most modern analytical laboratories are equipped with a GC/MS, many laboratories will be able to perform the analysis, and the sample load will be decreased. Second, GC introduction will provide improved sample introduction techniques that are more convenient and less prone to component discrimination. Additionally, the use of chromatography will provide information such as distribution of components as a function of boiling point and make possible the extension of the method to total fuel analysis by increasing the chromatographic resolution.

Sample Introduction

The utilization of GC/MS techniques can provide more up-to-date methods of sample introduction into the mass spectrometer. This is a direct advantage over the traditional ASTM methods, in which a heated glass inlet or microburette is employed, and the sample is introduced by effusion through a small orifice. A small fraction of the total sample is analyzed with the microburette technique, in order to minimize component discrimination that increases as the effusion of the sample into the mass spectrometer proceeds. The GC/MS technique will vaporize the entire sample and analyze the entire volume that has been loaded onto the column. Because of the possibility of component discrimination during sample loading onto the GC column, component discrimination must be evaluated as a function of injection port temperature and injection techniques.

Short Column GC/MS and Simulated Distillation

An important concept of this HTA method is that chromatographic resolution of the jet fuel components is not a matter of pursuit. Trehy et al. [*7*] have discussed some of the important instrumental considerations of normal GC/MS using short capillary columns. In the GC/MS/HTA application of short capillary columns, the GC is mainly used as a convenient device for sample introduction into the mass spectrometer. The short columns, along with providing short analysis times, provide the ability to observe the distribution profile of the eluting jet fuel as a function of column temperature. This concept is similar to that of simulated distillation that has been utilized by the petroleum industry for many years to provide a boiling point distribution of the hydrocarbon mixture.

Experimental

Materials

All chemicals were American Chemical Society (ACS) grade unless otherwise specified and were used as received without further purification. Standards were purchased from various vendors and were used as received without further purification. All jet fuel samples were obtained from Pratt & Whitney-Fuels Division, West Palm Beach, FL. The fuels were shipped at ambient temperatures without refrigeration in vials with screw-cap or crimped cap teflon-lined replaceable septa. After every use in the laboratory, the septa were replaced with new ones and samples were refrigerated at 5°C.

Apparatus

Mass Spectrometer—A Finnigan MAT TSQ45 gas chromatograph-triple-stage quadrupole mass spectrometer/data system was used in all GC/MS studies. An electron energy of 70 eV and an electron current of 300 μA were used. The mass spectrometer was mass calibrated with FC43 (perfluorotributylamine) in accordance with the instrument's specifications. The mass spectrometer was tuned, also with FC43, to obtain a specific ion intensity ratio of m/z 69 to m/z 219, and m/z 502 to m/z 220. These ratios were monitored to ensure reproducible relative abundances of the fragmentation pattern in regards to instrumental tuning alone. The low mass ratio is of greater importance in these studies since this is the general mass range in question. The mass spectrometer was scanned from 35 to 650 atomic mass units (amu) at a total scan time of 1.0 s. The electron multiplier was operated between 950 and 1000 V, with the conversion dynodes at $\pm$3000 V. The preamplifier gain was set at 10^8 V/A with the fine adjustment (zeroing) set for minimal acquisition of baseline noise. The mass spectrometer was interfaced to a Finnigan Model 9610 gas chromatograph. All GC operations were under computer control. The capillary GC columns were inserted directly into the ion source of the mass spectrometer through a heated interface. The ion source pressure was 0.3 to 0.6 torr depending on the column head pressure. The ion source temperature was 190°C.

Gas Chromatography—A 3-m DuraBond fused silica bonded phase open tubular (capillary) column (J&W, Rancho Nuevo, CA) with an inner diameter of 0.25 mm and a DB5 bonded phase (SE-54 equivalent or 5% phenyl-methyl silicone) of 1.0 μm thickness was operated at a head pressure of 54 kPa (8 psi) of helium carrier gas. Split-type injection was used with a split flow of 50 mL/min and a septum sweep of 2 mL/min. The injection port temperature was 250°C and the transfer line was 280°C. The injection port teflon lined septum was replaced after 30 to 40 injections and the injection port glass insert was checked frequently and cleaned whenever needed. The column temperature was programmed from 50 to 250°C at 15°C/min after a hold time of 1 min. Usually, the column would not require the full temperature to elute the entire jet fuel sample. Generally, 0.5 to 1.0 μL injections of neat jet fuel samples were made. The injection technique used was as follows in order to reduce injection error: the GC injection syringe (Hamilton) was thoroughly rinsed between samples with hexane, followed by the sample jet fuel of interest; the sample volume was pulled up from the needle volume into the syringe volume; the needle was inserted into the GC injector; a count of three allowed the needle to come to equilibrium temperature; the sample was injected at a rate of approximately 1 μL/s; this position was held for a count of three and the needle was removed. This method decreases injection discrimination due to uneven heating of the needle and components contained therein [*8,9*].

Other Gas Chromatographic Studies—Besides GC/MS, studies involved the application of GC with flame ionization detection. A Varian 3300 gas chromatograph was used with a flame ionization detector (FID). The same chromatographic conditions were used as just stated including the column head pressure. However, although the column head pressures were equivalent for the two systems, different flow rates resulted due to the different exit pressures. This instrument was used for the studies of component discrimination as a function of injection port temperatures and injection modes (split, splitless, and on-column). Acquisition of peak area, peak height, and retention time was obtained with an IBM 9000 minicomputer and the IBM CAPMC4 Chromatography Applications Program. Additionally, the FID system provided two advantages: a continuous mode of detection that resulted in more highly resolved chromatograms, and increased sensitivity (compared to full scan MS) that when combined with high resolution gas chromatography (longer columns, 30 m) would demonstrate the extreme complexity of the jet fuels.

Procedure

Ion Summation Chromatograms—The characteristic ion sums listed in Table 1 were plotted as separate mass chromatograms. These mass chromatograms, as shown in Fig. 1, are referred to as ion summation chromatograms and represent the chromatogram of each ion sum listed in Table 1. The area of each ion summation chromatogram was determined and normalized to 100%. The area percents were then calibrated with the ASTM inverse calibration matrices after determination of the average aromatic and average paraffinic carbon numbers. All these calculations were done automatically by a TURBO BASIC program written to perform such calculations.

Computer Hardware and Software—The GC/MS/HTA method required extensive computations once the mass spectral data had been obtained. A series of TURBO BASIC programs was written to completely automate these computations. Additionally, a terminal emulator program (Persoft, Smarterm 240) was utilized for the direct transfer of data from the INCOS data system to an IBM PC. The GC/MS/HTA TURBO BASIC program was written to calculate the average aromatic and average paraffinic carbon numbers as well as the volume percent results for each hydrocarbon type. An additional program was written to translate the emulator-captured data to a usable format for the GC/MS/HTA program. With these programs, the acquired data is never handled during data reduction.

Results and Discussion

GC Sample Introduction

With the application of gas chromatography (GC), a new method of sample introduction into the mass spectrometer was used for HTA. It was therefore important to evaluate the

TABLE 1—*The characteristic ion summations used for each hydrocarbon type in the hydrocarbon-type analysis.*

Hydrocarbon Type	Masses of Each Characteristic Ion Sum (Σ) (m/z)
Paraffins	Σ43 = 43 + 57 + 71 + 85 + 99
Monocycloparaffins	Σ41 = 41 + 55 + 69 + 83 + 97
Dicycloparaffins	Σ67 = 67 + 68 + 81 + 82 + 95 + 96
Alkylbenzenes	Σ77 = 77 + 78 + 79 + 91 + 92 + 105 + 106 + 119 + 120 + 133 + 134 + 147 + 148 + 161 + 162
Indanes/Tetralins	Σ103 = 103 + 104 + 117 + 118 + 131 + 132 + 145 + 146 + 159 + 160
Naphthalenes	Σ128 = 128 + 141 + 142 + 155

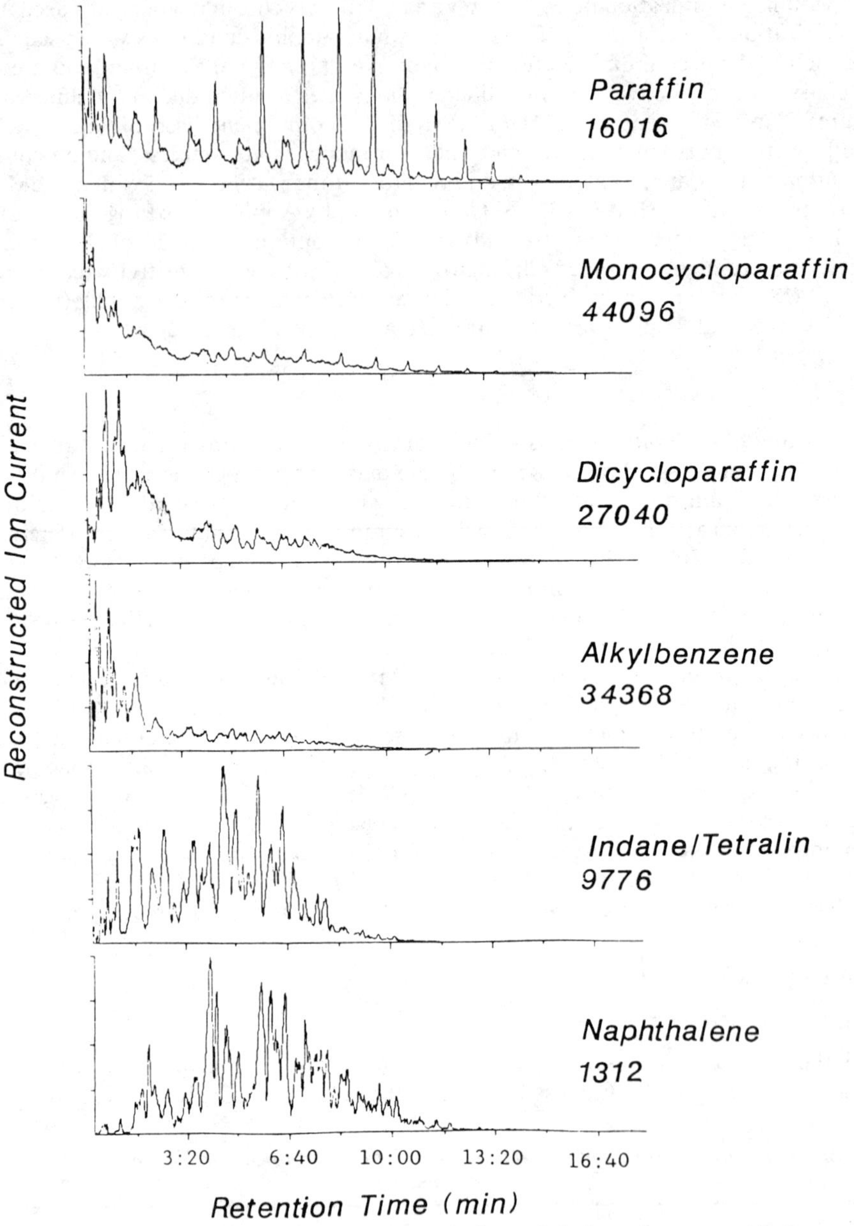

FIG. 1—*The ion summation chromatograms for each hydrocarbon type with the corresponding total area counts.*

specifics of sample discrimination with respect to this sample introduction technique. There are several injection techniques used in capillary gas chromatography. These are split and splitless, which depend on the flash vaporization of the sample within a heated glass-lined injection volume, and on-column injection that allows the sample to be loaded directly onto the capillary column at ambient temperatures. Specifics of these capillary GC injection techniques are discussed elsewhere [8,9]. An aromatic mixture containing C_{12}, C_{14}, C_{16}, and C_{18} straight chain alkylbenzene in equal concentrations was used for the evaluation of component discrimination. The effect of injection port temperature and injection technique was evaluated. Figures 2 and 3 illustrate the effect of injection port temperature on component discrimination for split and splitless injections, respectively, with flame ionization detection. In both cases, the ratios of peak areas were normalized to the C_{14}/C_{12} ratio at the 280°C injection port temperature to help illustrate the matter of discrimination.

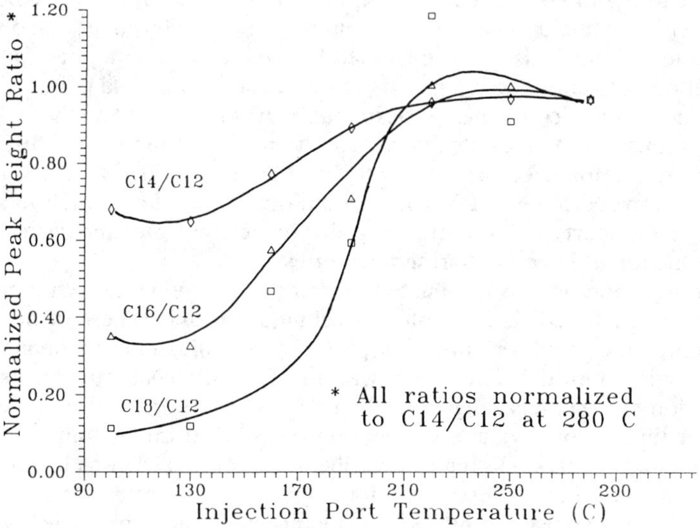

FIG. 2—*Plot of injection port temperature versus area ratios of C_{12}, C_{14}, C_{16}, and C_{18} alkylbenzenes for splitless injections on a 3-m DB5 capillary column, with FID detection.*

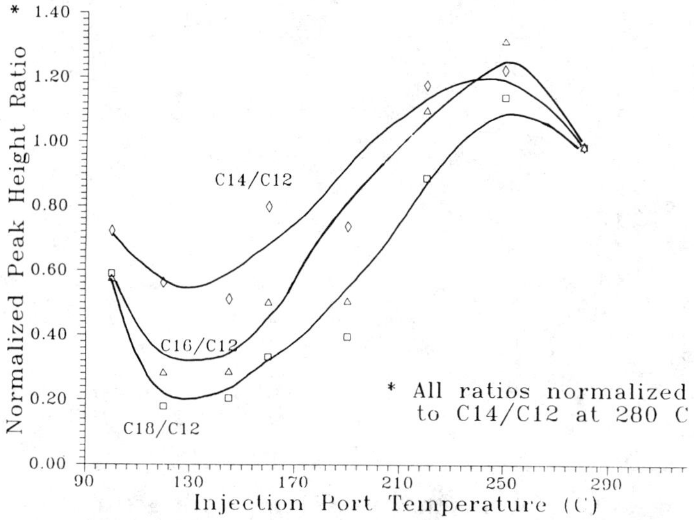

FIG. 3—*Plot of injection port temperature versus area ratios of C_{12}, C_{14}, C_{16}, and C_{18} alkylbenzenes for split injections on a 3-m DB5 capillary column, with FID detection.*

Both cases indicate component discrimination taking place at lower temperatures by the ratios being lower. This was due to less efficient vaporization of the heavier components at lower temperatures, as was expected. As the temperature was raised, the ratios increased and leveled off, indicating a lack of discrimination and the absence of component decomposition at higher injection port temperatures. A larger degree of discrimination takes place for the higher carbon number alkylbenzenes, as indicated by the larger ratio spread between the low and high injection port temperatures. The splitless technique more clearly illustrates these trends. A minimum injection port temperature of 250°C should therefore be maintained to avoid component discrimination. These results also demonstrate the improvement in reproducibility obtained using splitless injections over that obtained with split injection technique. This is demonstrated by the increased amount of scatter within the split injection data points compared to the splitless injection data points of Figs. 2 and 3. A similar evaluation of component discrimination was performed on the GC/MS instrument. Figure 4 illustrates the results that are similar to those from the GC/FID technique. However, discrimination effects are less with GC/MS, as indicated by the smaller spread of area ratios compared to GC/FID. The vacuum outlet conditions result in a large pressure drop across the short capillary column, giving higher flow rates and decreased discrimination as a function of injection port temperatures.

On-column injection was also evaluated in terms of component discrimination. It has already been demonstrated that on-column techniques provide the best precision when compared to the other modes of injection [*10*]. Figure 5 illustrates component discrimination as a function of initial column temperature. The initial column temperature in on-column injection corresponds roughly to the injection port temperature in split/splitless injections. For this reason, evaluations were not possible at initial column temperatures above 120°C, since the early eluting C_{12} alkylbenzene component was lost in the solvent front. The results indicated no discrimination, as would be expected for on-column injections. However, there appears to be a small enhancement effect for the less volatile components at lower initial column temperatures. This cannot be explained at the present time.

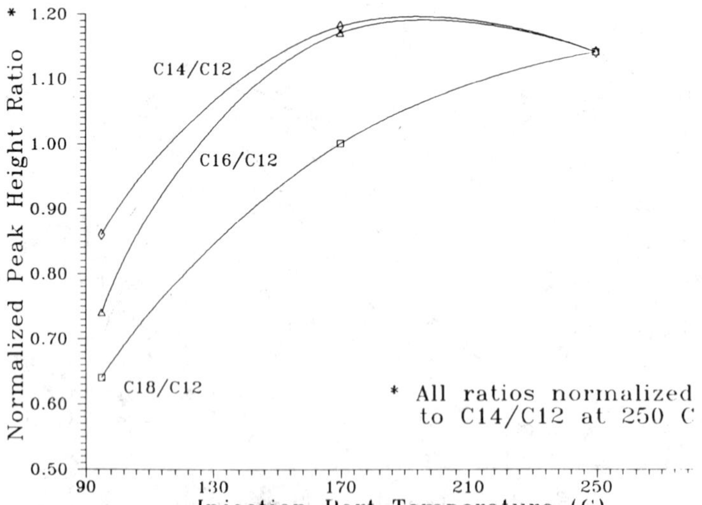

FIG. 4—*Plot of injection port temperature versus area ratios of C_{12}, C_{14}, C_{16}, and C_{18} alkylbenzenes for split injections on a 3-m DB5 capillary column and MS detection.*

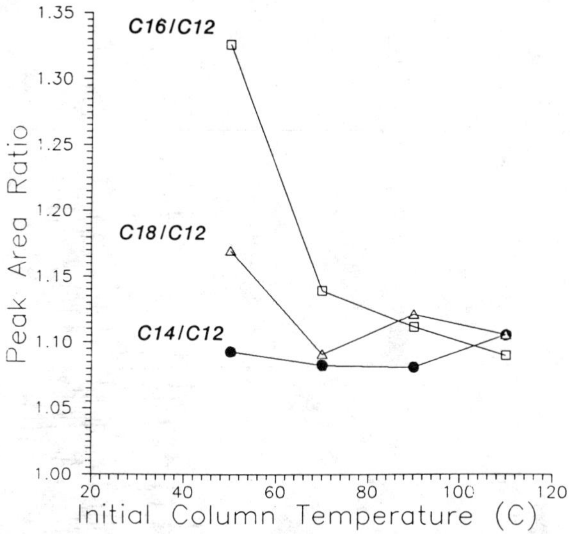

FIG. 5—*Plot of initial column temperature versus area ratios for C_{12}, C_{14}, C_{16}, and C_{18} alkylbenzenes for on-column injections on a 3-m DB5 column and FID detection.*

Table 2 compares splitless and on-column injection techniques for accuracy and precision. A mixture containing known weight ratios was prepared and analyzed with GC/FID. The FID response is proportional to the mass flux of analyte, and therefore direct correlations may be made between weight ratios and area ratios. Results for both splitless and on-column injections were high compared to the actual weight ratio, which may be due to some systematic error. The splitless results agreed more closely with the weight ratios, while the precision of the on-column results was somewhat better.

These results indicate that, as long as the minimum injection port temperature is maintained, component discrimination is greatly reduced if not eliminated. Also, better reproducibility is obtained using splitless injections.

Analytical Figures of Merit

The limits of detection that are determined here do not reflect the limits that would actually be obtained if individual components were to be determined. Figure 6 illustrates

TABLE 2—*Comparisons of accuracy and precision of area ratios of various components between splitless and on-column injection techniques.*

	Butylbenzene / Decane	Hexamethylbenzene / Decane	*t*-Stilbene / Decane
Actual weight ratio	0.785	0.280	0.374
Splitless area ratio	0.894(0.005)	0.30(0.01)	0.41(0.03)
%RSD	0.54	4.32	7.24
On-column area ratio	0.905(0.004)	0.330(0.009)	0.49(0.01)
%RSD	0.44	2.82	2.02

NOTE—Number of replicates $n = 3$; and number in parenthesis indicates standard deviation.

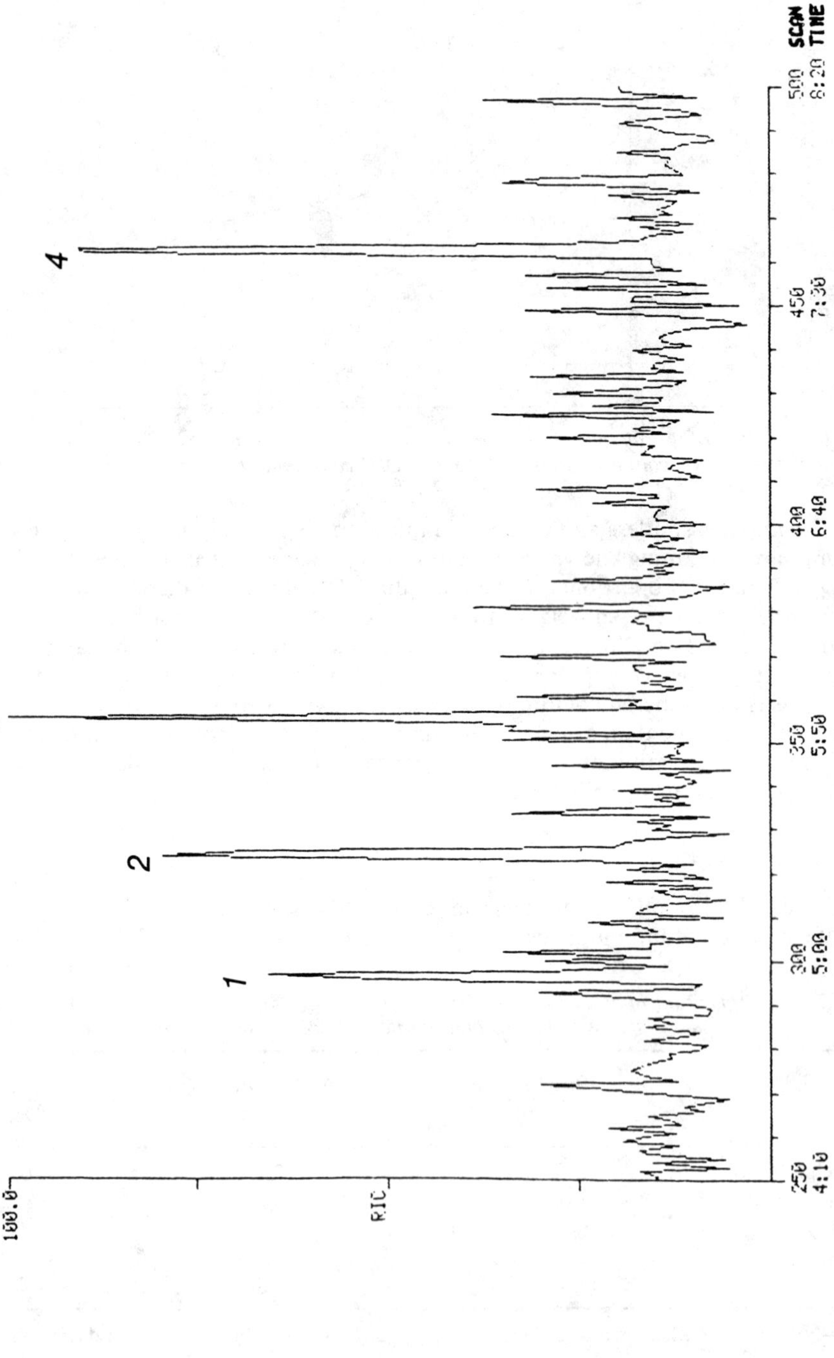

FIG. 6—*Chromatogram of nonane (1), cyclooctane (2), propylbenzene (3), and decalin (4) at 0.5% by volume each using 0.4 µL split injection on a 3-m column.*

the reconstructed ion current (RIC) full-scan mass spectra of a four-component mixture at 0.5% by volume each, which corresponds to approximately 0.4 ng injected on the column. The signal-to-noise in the RIC was approximately 3, which represents the limit of detection for the HTA method. The sensitivities of the four components are given by the slope of the calibration curves illustrated in Fig. 7. These sensitivities were determined in order to compare the results to the actual ASTM sensitivities (Table 5 of ASTM D 2789-86). Table 3 compares the sensitivities (slopes of the response curves in Fig. 7) to the sensitivities of Table 5 of ASTM D 2789-86), after normalization to the C9 paraffin (nonane) sensitivities. The results indicate that similar trends in sensitivities for the four components were observed for the two methods, although the propylbenzene sensitivity was significantly greater in the ASTM results.

The precision of the HTA method was determined by replicate analyses of several jet fuel samples. The precision was dependent upon the jet fuel analyzed or the composition of the fuels. The precision of the HTA method is summarized in Table 4, in which the average percent relative standard deviation is listed by hydrocarbon type for all fuels analyzed in replicate. The high average relative standard deviations were a result of that particular percent by volume being very low. The dashed lines indicate a consistent result of zero percent by volume.

Comparisons of ASTM and GC/MS/HTA Results

A number of real jet fuel samples were analyzed by the newly developed GC/MS/HTA method. The results of the analyses of these fuels were compared to results from the long-

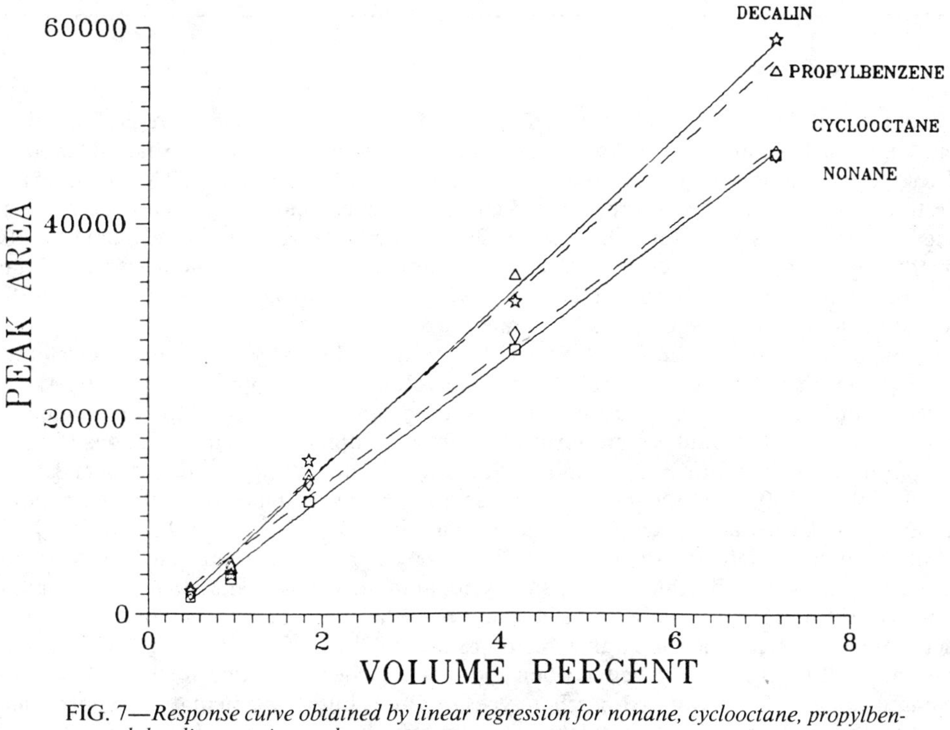

FIG. 7—*Response curve obtained by linear regression for nonane, cyclooctane, propylbenzene, and decalin at various volume percentages.*

TABLE 3—*Comparison of sensitivities of ASTM and GC/MS/HTA results for a four-component mixture normalized to the ASTM sensitivity of nonane (response per percent by volume).*

Component	ASTM Sensitivity	GC/MS/HTA Sensitivity
Decalin	1.99	2.15
Propylbenzene	2.52	2.06
Cyclooctane	1.72	1.74
Nonane	1.72	1.72

TABLE 4—*The average relative standard deviation for the GC/MS/HTA of all replicated analyses and separate jet fuels.*

Hydrocarbon Type	Average of all Fuels	JP-4 2455	JP-7	JP-8X 2414	JP-8X 2429
Paraffin	3.0	2.1	1.2	7.1	1.7
Monocycloparaffin	3.8	2.5	0.6	6.8	5.3
Dicycloparaffin	5.6	18.1	1.2	1.4	1.8
Alkylbenzene	7.6	3.8	18.8	2.5	5.4
Indane/tetralin	5.0	10.7	. . .	2.0	2.4
Naphthalene	7.7	0.0	. . .	8.7	6.7
		AVERAGE CARBON NUMBER			
Aromatic	0.5	0.5	0.9	0.5	0.2
Paraffin	0.3	0.7	0.3	0.1	0.0

established ASTM D 2789-86 method. The ASTM D 2789-86 analyses were performed at the National Institute of Petroleum and Energy Research (NIPER) in Ada, Oklahoma. These comparisons would provide information on the validity of the ASTM inverse calibration matrices with GC sample introduction and quadrupole mass analysis. Table 5 shows these comparisons for five jet fuels. Several methods were used to evaluate trends in the comparison versus composition, but none could be found. The accuracy of the ASTM method has not been evaluated; therefore, the comparison does not necessarily reflect the accuracy of the GC/MS/HTA method.

Table 6 demonstrates the comparisons of several synthetically prepared jet fuels. The "recipe" results were calculated from the known volumes of the components added. Again these comparisons do not necessarily represent accuracy of the GC/MS/HTA method, since the synthetic mixtures were formulated from complex petroleum fractions (Exxon Corporation) such as Isopar C, G, and M (isoparaffinic and paraffinic mixtures), and Aromatic 100 and 150 (aromatic mixtures). These mixtures were obtained by simple fractionation and bulk separation techniques, and were not intended for analytical standards. Two of these mixtures were analyzed by GC/MS for specific group components. The Isopar M mixture contained significant amounts of cycloparaffinic components, with trace aromatics. The Aromatic 150 contained indanes and naphthalenes; the paraffin content was not determined. The monocycloparaffin recipe results of the olefin-containing fuels (Grand Mix and All Groups) agree better when the olefin content is added to the monocycloparaffin content. These two hydrocarbon types produce similar fragmentation patterns that tend to equate the two types under the HTA method.

TABLE 5—*Comparison of GC/MS/HTA and ASTM D 2789-86 results on various fuels in percent by volume.*

Fuel	Paraffin	Monocyclo-paraffin	Dicyclo-paraffin	Alkyl-benzene	Indane-tetralin	Naphthalene	Aromatic Carbon Number	Paraffin Carbon Number
JP4 2455								
GC/MS/HTA	61.5(0.2)	18.5(0.8)	3(1)	14.8(0.9)	2.0(0.4)	0	9.06(0.07)	9.50(0.08)
ASTM D 2789-86	70.8	18.0	2.1	7.6	1.4	0.1	8.69	8.01
JP8X 2429								
GC/MS/HTA	0	3.2(0.8)	68(2)	6.6(0.7)	21.4(0.8)	1.3(0.3)	8.0(0.1)	9.01(0.01)
ASTM D 2789-86	1.3	21.5	56.0	7.2	13.1	0.9	8.14	8.91
JP8X 2414								
GC/MS/HTA	7.0(0.5)	10.3(0.7)	35.0(0.5)	15.7(0.4)	29.7(0.6)	2.3(0.2)	9.87(0.05)	9.40(0.01)
ASTM D 2789-86	14.2	27.0	26.6	15.3	14.5	2.3	9.18	9.46
JP8X SIMULATED								
GC/MS/HTA	11.5(0.2)	17.0(0.9)	54.9(0.9)	7.4(0.4)	8.5(0.2)	0.6(0.04)	8.53(0.02)	9.00(0)
ASTM D 2789-86	12.1	23.1	58.7	2.9	4.1	0.2	7.65	8.69
JETA 2532								
GC/MS/HTA	18.1	38.1	33.4	5.7	4.6	0	9.32	9.30
ASTM D 2789-86	27.4	44.7	20.8	4.9	2.1	0.2	8.41	8.86

NOTE—numbers in parentheses indicate the standard deviation, $n = 3$.

TABLE 6—*Comparison of GC/MS/HTA results and synthetic fuel recipes for various synthetic fuels in percent by volume.*

Fuel	Paraffin	Monocyclo-paraffin	Dicyclo-paraffin	Alkyl-benzene	Indane-tetralin	Naphthalene	Olefins
JP8X simulated							
GC/MS/HTA	11.5	17.5	54.9	7.4	8.5	0.6	. . .
Recipe	12	20	60	(combined unsaturates 8%)			. . .
High Aromatic							
GC/MS/HTA	13.5	4.0	3.8	51.0	13.9	13.8	. . .
Recipe	30	(combined 10%)		45	10	5	. . .
Grand Mix							
GC/MS/HTA	23.0	14.0	14.8	30.3	0.1	17.5	. . .
Recipe	40	10	20	20	0.0	5	5
All Groups							
GC/MS/HTA	36.6	17.2	2.9	26.7	4.7	11.8	. . .
Recipe	60	(combined 10%)		17	3	5	5

Conclusions

A new HTA method was developed for the analysis of jet fuels using GC/MS and the pre-existing ASTM D 2789-86 inverse calibration matrix. The GC/MS/HTA method was characterized by the comparison of the results of several test fuels to the actual ASTM D 2789-86 results. The GC/MS/HTA results are reproducible and relatively consistent with ASTM D 2789-86 results. Although the actual accuracy of the ASTM results have yet to be evaluated, the ASTM method still represents the industry standard. Additional confirmation experiments included the analysis of various synthetic fuel mixtures (pseudo-fuels) that were compared to the actual "recipe results" of these mixtures.

The presence of component discrimination with GC sample introduction was investigated as a function of injection port temperature and injection mode. Results demonstrate that as long as a minimum injection port temperature is maintained, component discrimination is eliminated.

The GC/MS/HTA method will make available to many more laboratories the ability to perform such analyses. On that matter, multiple analyses and inter-laboratory comparisons of results will help improve the method and increase the accuracy and reproducibility. Additionally on this matter, new calibration schemes may be developed that will greatly improve the overall accuracy of the technique. This includes sliding scale methods that depend on GC separations. In our laboratory, it is believed that sliding-scale inverse calibration matrixes based on Kovats indices will be the most logical means of obtaining this goal.

References

[*1*] Brown, R. A., *Analytical Chemistry,* Vol. 23, No. 3, March 1951, pp. 430–437.
[*2*] Lumpkin, H. E., Thomas, B. W., and Elliott, A., *Analytical Chemistry,* Vol. 24, No. 9, Sept. 1952, pp. 1389–1391.
[*3*] Hood, A., *Analytical Chemistry,* Vol. 30, No. 9, July 1958, pp. 1218–1220.
[*4*] O'Neal, M. J., Jr., and Wier, T. P., Jr., *Analytical Chemistry,* Vol. 23, No. 6, June 1951, pp. 830–843.
[*5*] Teeter, R. M. and Doty, W. R., *Reviews of Scientific Instruments,* Vol. 37, 1966, pp. 792–793.
[*6*] Bieman, K., *Mass Spectrometry: Organic Chemical Applications,* McGraw Hill, New York, 1962.
[*7*] Trehy, M. L., Yost, R. A., and Dorsey, J. G., *Analytical Chemistry,* Vol. 58, No. 1, Jan. 1986, pp. 14–19.
[*8*] Grob, K. and Grob, G., *Chromatographia,* Vol. 5, 1972, p. 3.
[*9*] Grob, K. and Grob, K., Jr., *Journal of Chromatography,* Vol. 94, 1974, p. 53.
[*10*] McMahon, D. H., *Journal of Chromatographic Science,* Vol. 23, 1985, p. 137.

David J. Miller[1] *and Steven B. Hawthorne*[1]

Chemical Ionization-Proton Exchange Mass Spectrometric Identification of Alkyl Benzenes, Tetralins, and Indans in a Coal-Derived Jet Fuel

REFERENCE: Miller, D. J. and Hawthorne, S. B., **"Chemical Ionization-Proton Exchange Mass Spectrometric Identification of Alkyl Benzenes, Tetralins, and Indans in a Coal-Derived Jet Fuel,"** *Novel Techniques in Fossil Fuel Mass Spectrometry, ASTM STP 1019,* T. R. Ashe and K. V. Wood, Eds., American Society for Testing and Materials, Philadelphia, 1989, pp. 38–46.

ABSTRACT: The lack of appropriate standards and the similarity of electron impact (EI) ionization mass spectra shown by isomeric alkyl benzenes, indans, and tetralins often makes the identification of specific isomers impossible. Chemical ionization-proton exchange (CIPE) mass spectrometry using methanol-d_1 as the reagent gas aids in their identification by determining the number of unsubstituted aromatic positions. After ionization of the aromatic ring (that is, addition of D^+), only the aromatic hydrogens undergo exchange for deuterium. The number of deuterium exchanges observed allows one to assign the degree of alkyl substitution on the aromatic ring. The CIPE spectra also provide information on the position of alkyl substitution. For example, ortho and para disubstituted alkylbenzenes show a base peak (100% relative intensity) at the mass representing the exchange of all the aromatic protons compared to an intensity of 30 to 40% for meta-disubstituted benzenes. Gas chromatographic/mass spectrometric analysis of the aromatic fraction of a coal-derived jet fuel using traditional EI ionization showed that this fraction (57% by weight) was composed primarily of alkyl substituted benzenes, indans, and tetralins, but few specific isomers could be identified. The CIPE analysis allowed additional classification of approximately 50 individual alkylbenzene, indan, and tetralin isomers based on the degree of aromatic substitution.

KEY WORDS: chemical ionization-proton exchange, mass spectrometry, coal-derived jet fuel, alkylbenzenes, alkylindans, alkyltetralins, isomer identification, fossil fuels

The chemical structure of future jet fuels was the subject of a recent symposium sponsored by the Division of Petroleum Chemistry, Inc., American Chemical Society. Papers presented described the production of jet fuels from coal syncrudes [*1*], tar and oil sands [*2,3*], oil shale [*4*], and gasifier tar oil [*5*]. Jet fuels that are derived from nonpetroleum feedstocks will need to be thoroughly characterized in order to determine their suitability as replacements for current petroleum-based fuels. The U.S. Air Force has funded numerous studies relating to the analysis of aircraft fuels. These studies cover the range of organic analytical techniques from spectroscopy to chromatography [*6–9*]. The use of capillary gas chromatography (GC) with flame ionization (FID) or mass spectrometric (MS) detection provides an extremely powerful tool for the characterization of complex mixtures such as jet fuel [*10,11*]. Retention indices of standard compounds can sometimes be used for ten-

[1] Research associate and research associate, repsectively, University of North Dakota, Energy and Mineral Research Center, Grand Forks, ND, 58202.

tative compound identification and to establish elution order for isomeric compounds. Although there are numerous published lists of retention indices, standards for many of the compounds found in coal-derived fuels are not available. Mass spectrometry provides information that can establish the identity of unknown sample components by matching electron impact (EI) ionization fragmentation patterns with standard spectra. Unfortunately, many isomeric compounds have nearly identical EI mass spectra. This limits the identification of many species and allows only the assignment of compound class, for example, C_3 benzene.

Chemical ionization–proton exchange (CIPE) mass spectrometry can be used to provide additional structural information by determining the number of protons bonded to an aromatic ring [*12*]. Aromatic species such as benzenes, indans, and tetralins undergo ionization by the addition of D^+, followed by the exchange of aromatic protons for deuterium. These exchanges can be used to determine the number of alkyl substitutions on the aromatic ring. For example, a propylbenzene will show an M + 7 ion (ionization by D^+ and five exchanges) while a trimethylbenzene will show an M + 5 ion (ionization by D^+ and three exchanges). Since indans and tetralins are disubstituted benzenes, the maximum number of aromatic proton exchanges is four. The information provided by CIPE for indans and tetralins allows one to determine whether additional alkyl substitution is on the aromatic or the aliphatic ring.

In this study, CIPE was used to aid in the identification of alkylbenzenes, alkylindans, and alkyltetralins in a coal-derived jet fuel. The CIPE spectra for 24 alkylindan and alkyltetralin isomers are reported to augment spectra previously reported for alkylbenzenes [*12*].

Experimental

Sample

A sample of raw tar oil from the Great Plains Gasification Plant (Beulah, North Dakota) was hydrotreated over a cobalt molybdenum catalyst (Harshaw Chemical Company) for 60 min at 410°C in a 3.78 L (1-gal) autoclave [*5*]. The hydrogenated tar oil was separated using activated silica-gel-column chromatography to obtain three fractions; aliphatics, aromatics and polars [*5*]. This study deals with the aromatic fraction from this separation.

Gas Chromatography/Mass Spectrometry

A Hewlett-Packard Model 5985B GC/MS equipped with a dual electron impact (EI)/chemical ionization (CI) source was used to obtain all of the spectra reported in this study. When the instrument was operated in the CI mode, the methanol-d_1 reagent gas was introduced directly into the mass spectrometer ion source through a 1.58 mm (1/16 in.) outside diameter stainless steel tube running through the heated transfer block co-linearly with the chromatographic column. Source parameters and reagent gas pressure were optimized by introducing *m*-diisopropylbenzene at a constant rate into the source through the direct insertion probe and maximizing the intensity of the M + 6 ion that is formed from the addition of D^+ and the exchange of all four aromatic hydrogens for deuterium [*12*]. The reagent gas pressure was 0.2 torr. The column used for gas chromatographic separations was a 60 m × 0.25 mm inside diameter (0.25 μm film thickness) DB-5 fused-silica capillary column (J & W Scientific, Folsom, CA). The column oven temperature was ramped from 60 to 320°C at 6°C/min.

Results and Discussion

The use of CIPE to determine the degree of aromatic substitution for alkylbenzene, alkylindan, and alkyltetralin isomers is shown in Fig. 1. The number of unsubstituted positions on a benzene ring can be determined by the formula

$$X = M_2 - M_1 - 2$$

where X is the number of unsubstituted aromatic positions, M_1 is the molecular weight of the alkylbenzene, alkylindan, or alkyltetralin, M_2 is the mass corresponding to the maximum exchange of the aromatic protons, and the number 2 is subtracted to correct for the ionization of the aromatic ring by D^+ during chemical ionization [*12*]. For example, *n*-hexylbenzene (molecular weight = 162) shows ionization by D^+ and five H/D exchanges (*m/z* 169), *p*-diisopropylbenzene shows ionization by D^+ and four H/D exchanges (*m/z* 168), and hexamethylbenzene shows only ionization by D^+, (*m/z* 164, Fig. 1*a*).

Since indans and tetralins are disubstituted benzenes, the maximum number of aromatic protons available for exchange is four and the maximum shift in molecular weight is six (two for ionization by D^+ and four for exchange of all aromatic protons). Under the conditions used in this study, no exchange of aliphatic protons was observed for either alkylbenzenes, alkylindans, or alkyltetralins. Indans and tetralins that are substituted on the aliphatic ring yield intense M + 6 ions. Examples of this are shown in Fig. 1*b* and *c* (1,1-dimethylindan and 1,1-dimethyltetralin, respectively). Substitution on both aromatic and aliphatic carbons or only on aromatic carbons is apparent by a decrease in the number of H/D exchanges. For example, 1,1-dimethyltetralin has an intense ion at *m/z* 166 (ionization by D^+ and four H/D exchanges); 1,8-dimethyltetralin has an intense ion at *m/z* 165 (ionization by D^+ and three H/D exchanges); and 5,8-dimethyltetralin has an intense ion at *m/z* 164 (ionization by D^+ and two H/D exchanges), Fig. 1*c*. The 26 CIPE spectra for alkylbenzenes that have been previously reported [*12*] and all of the alkylindan and alkyltetralin standards used in this study gave predictable CIPE spectra that followed the formula $X = M_2 - M_1 - 2$. The CIPE mass spectra for 24 alkylindans and alkyltetralins are given in Table 1.

CIPE spectra also contain information on the position of the substituents on the benzene ring. Alkylbenzene isomers that are ortho or para disubstituted show the greatest spectral intensity at the mass representing the exchange of all aromatic protons (that is, the M + 6 peak), while meta disubstituted alkylbenzenes show an intensity of 30 to 40% at the M + 6 peak [*12*]. This is demonstrated in Fig. 2*a* by the CIPE spectra of the three diethylbenzene isomers. Several of the standard alkylindan compounds also exhibit different relative intensities in the molecular ion region depending on the position of the alkyl substitution (Fig. 2*b*). If the compound is disubstituted on aromatic carbons adjacent to the aliphatic ring (carbon-4 and 7), the most intense ions are those corresponding to ionization by D^+ and exchange of one or two aromatic protons (M + 3 or M + 4). If one of the substituents is on an aromatic carbon adjacent to the aliphatic ring and one of the substituents is on an aromatic carbon once removed from the aliphatic ring (carbon 5 or 6), the most intense ion is at M + 2 (ionization by D^+) with less intense ions for the exchange of the two aromatic protons. If both substituents are on carbons once removed from the aliphatic ring (carbons 5 and 6), the most intense ion is at M + 4 (ionization and the H/D exchanges). Alkyltetralins with the same number of aromatic substituents also show differences in intensities depending on the position of the substitution. While the EI ionization spectra of these compounds are often indistinguishable, the relationship between the exchange intensity and the position of aromatic substitution shown in the CIPE spectra

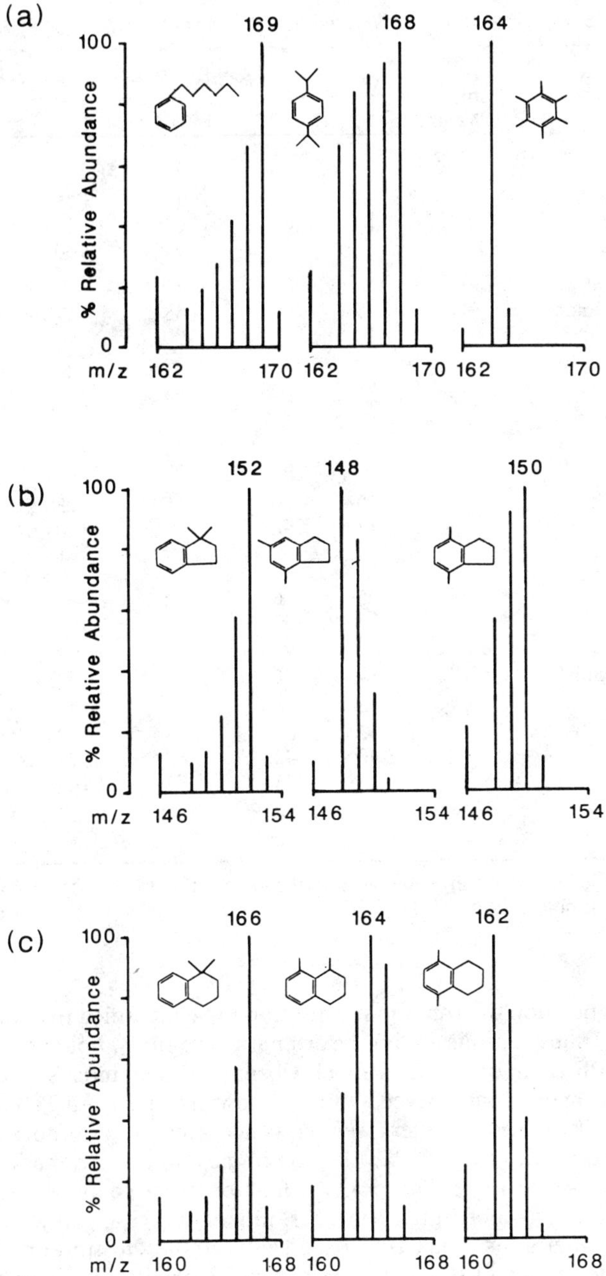

FIG. 1—*Chemical ionization-proton exchange mass spectra. The species shown from left to right are* (a) n-*hexylbenzene*, p-*diisopropylbenzene, and hexamethylbenzene;* (b) *1,1-dimethylindan, 4,7-dimethylindan, and 4,8-dimethylindan; and* (c) *1,1-dimethyltetralin, 1,8-dimethyltetralin, and 5,8-dimethyltetralin.*

TABLE 1—*CIPE mass spectra of alkylindan and alkyltetralin isomers with CH_3OD reagent gas.*

Compound	Molecular Weight	Relative Abundance, %[a]					
		M	(M + 2)	(M + 3)	(M + 4)	(M + 5)	(M + 6)
5-methylindan	132	38	56	77	100	85	
1,1-dimethylindan	146	13	10	13	25	58	100
4,6-dimethylindan		10	100	83	33		
4,7-dimethylindan		21	56	92	100		
1,4,7-trimethylindan	160	13	65	98	100		
1,1,4,6-tetramethylindan	174	5	100	58	17		
1,1,4,7-tetramethylindan		10	69	100	90		
1,1,5,6-tetramethylindan		10	25	54	100		
1,1,3,3,5-pentamethylindan	188	25	60	95	100	70	
5-methyltetralin	146	29	40	71	100	90	
6-methyltetralin		42	98	100	85	50	
1,1-dimethyltetralin	160	15	10	15	25	58	100
1,8-dimethyltetralin		17	48	75	100	90	
5,7-dimethyltetralin		6	100	69	21		
5,8-dimethyltetralin		25	100	75	40		
2-ethyltetralin		25	6	12	23	56	100
5-ethyltetralin		23	50	83	100	79	
6-ethyltetralin		29	63	90	100	88	
1,5,8-trimethyltetralin	174	17	100	67	27		
1,1-diethyltetralin	188	4	17	23	35	63	100
2,2,5,7-tetramethyltetralin		2	100	65	18		
2-t-butyltetralin		40	42	67	75	90	100
6-n-butyltetralin		27	46	77	100	90	
6,7-diethyl-1,1,4,4-tetramethyltetralin	244	13	100	56	17		

[a] All CIPE mass spectra were generated as described in the text from approximately 150 ng injected into the chromatographic column of the GC/MS. Ion abundances have been corrected for the contribution of ^{13}C isotope.

should allow the position of aromatic substitution to be identified in addition to determining whether alkyl substitutions are on the aromatic or on the aliphatic ring.

The use of CIPE technique to identify alkylbenzenes, alkylindans, and alkyltetralins in an aromatic fraction of a coal-derived jet fuel is shown in Fig. 3 and Table 2. Alkylbenzene, alkylindan, and alkyltetralin isomers were classified according to molecular weight using the M and M + 2 masses from CIPE. Electron impact, ionization mass spectra were used to confirm these classifications. The aromatic fraction of the coal-derived jet fuel consisted primarily of C_1-C_5 alkyl substituted benzenes, indan, and C_1-C_3 alkylindans, and tetralin and C_1 and C_2 alkyltetralins. The CIPE spectra allowed the substitution of 26 alkylbenzene isomers to be determined. Ortho or para versus meta substitution was also determined for six disubstituted alkylbenzenes. The CIPE spectra were also used to determine whether the substitution was on the aromatic or aliphatic ring for 25 alkylindan and alkyltetralin isomers. For example, Peak 28 showed an M + 6 ion, demonstrating that the methyl group was on the aliphatic ring, and Peak 29 showed an M + 5 ion, indicating that the methyl group was on the aromatic ring. The identities of the species with specific names in Table 2 were confirmed by comparing their CIPE spectra and retention indices with those of

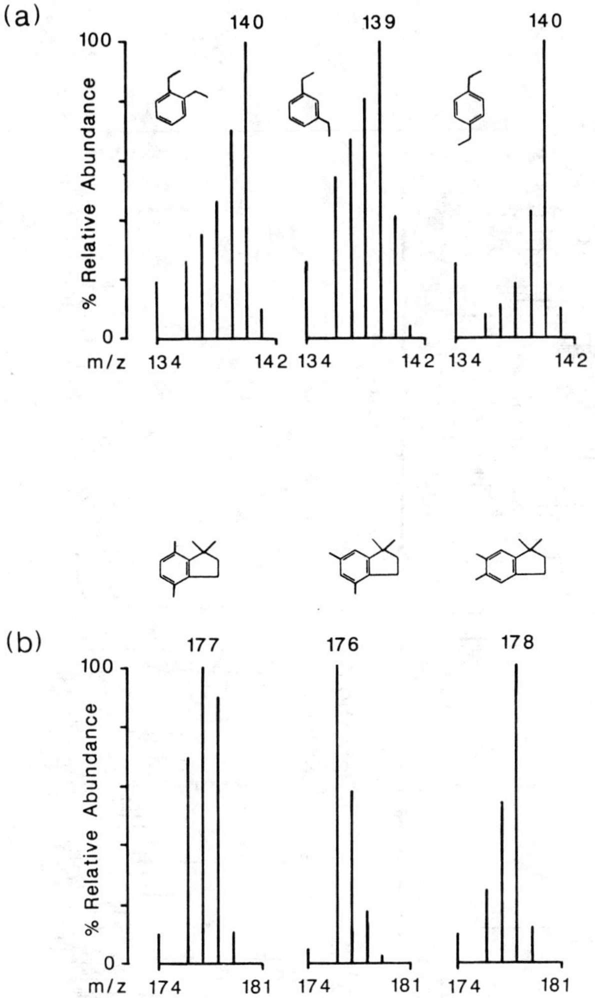

FIG. 2—*Effect of position of substitution on CIPE spectra for (from left to right)* (a) *o-diethylbenzene,* m-*diethylbenzene, and* p-*diethylbenzene and* (b) *1,1,4,7-tetramethylindan, 1,1,4,6-tetramethylindan, and 1,1,5,6-tetramethylindan.*

standards. Several higher weight aromatic species (Peaks 51 to 56) that apparently did not undergo complete hydrogenation, with subsequent ring opening to alkylbenzenes, alkylindans, and alkyltetralins, were also identified in the sample.

Conclusions

The analysis of the complex hydrocarbon mixture found in coal-derived jet fuel demonstrates the ability of CIPE mass spectrometry to provide structural information about alkylbenzene, alkylindan, and alkyltetralin isomers that is generally unavailable from EI spectra. The number of unsubstituted aromatic carbons can easily be determined without standard spectra or the study of fragmentation patterns, making CIPE ideal for the analysis

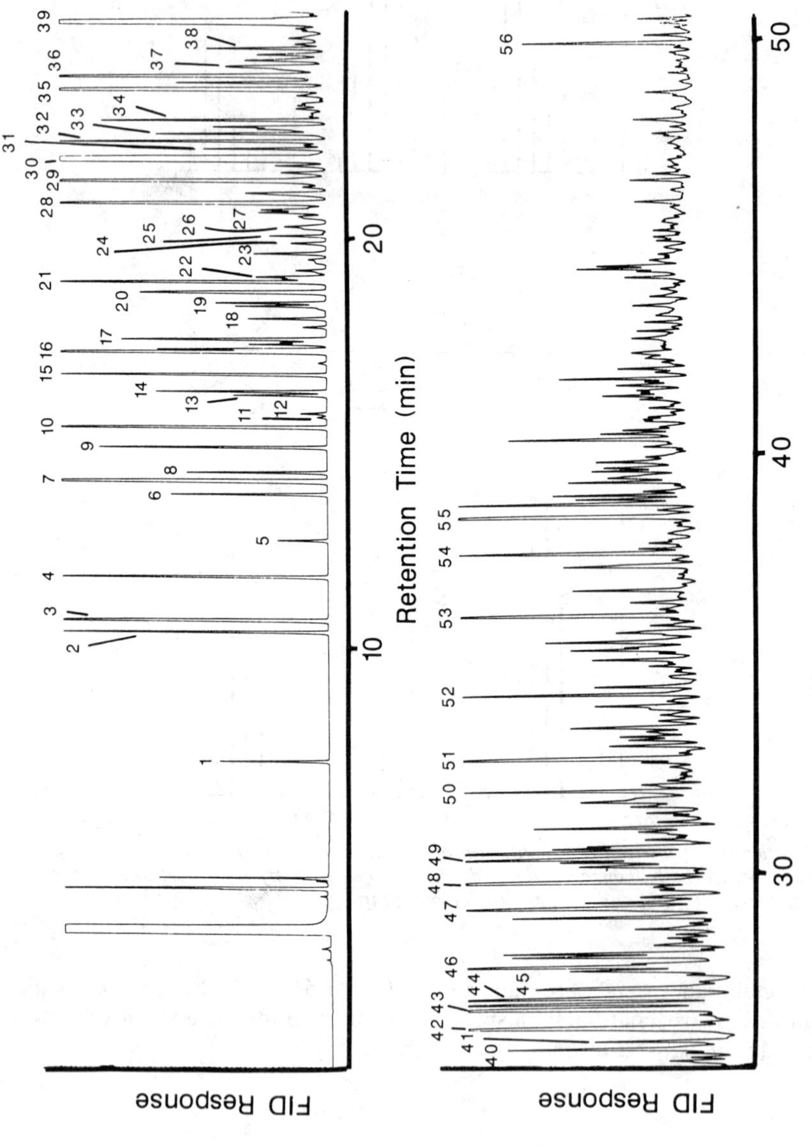

FIG. 3—*Capillary GC/FID chromatogram of organic species found in a coal-derived jet fuel.*

TABLE 2—*Identification of numbered peaks from Fig. 3.*

Peak No.	Species	Number of Substitutions on the Aromatic Ring[a]
1	toluene	1
2	ethylbenzene	1
3	*m*- and *p*-xylene	2
4	*o*-xylene	2
5	iso-propylbenzene	1
6	n-propylbenzene	1
7	*m*-ethylmethylbenzene and *o,p*-ethylmethylbenzene[b]	2
8	trimethylbenzene	3
9	*o,p*-methylethylbenzene[b]	2
10	trimethylbenzene	3
11,12,13	methylpropyl- or diethylbenzene	2
14	trimethylbenzene	3
15	indan	0
16,17	dimethylethylbenzene	3
18	methylpropyl- or diethylbenzene	2
19	dimethylethylbenzene	3
20	methylindan	0
21	dimethylethylbenzene	3
22	dimethylethylbenzene	3
23	tetramethylbenzene	4
24	dimethylethylbenzene	3
25	methylbutyl- or ethylpropylbenzene	2
26	dimethylethylbenzene	3
27	dimethyl- or ethylindan	0
28	methylindan	0
29	5-methylindan	1
30	tetralin	0
31	methyldiethyl- or dimethylpropyl benzene	3
32,33,34,37,42	dimethylindan or ethylindan or methyltetralin	1
35,36	dimethylindan or ethylindan or methyltetralin	0
38	4,6-dimethylindan	2
39	6-methyltetralin	1
40,41	trimethylindan or methylethylindan or dimethyltetralin	2
43,44	trimethylindan or methylethylindan or dimethyltetralin or ethyltetralin	1
45	1,4,7-trimethylindan	2
46	2,6- and 2,7-dimethyltetralin	1
47	1,8-dimethyltetralin	1
48	6-ethyltetralin	1
49	5,7-dimethyltetralin	2
50	5,8-dimethyltetralin	2
51	acenaphthene	
52	dibenzofuran	
53	fluorene	
54,55	octahydrophenanthrene/octahydroanthracene	
56	pyrene	

[a] For alkylindans and alkyltetralins, the number of aromatic substitutions given does not include the aliphatic ring.

[b] The CIPE spectra could not be used to differentiate between ortho and para isomers.

of highly complex samples. A study of the relative intensities of a particular specie's CIPE spectra also allows identification of some positional isomers and is particularly useful for distinguishing ortho and para from meta disubstituted alkylbenzenes.

References

[*1*] Sullivan, R. F., *Preprints,* Division of Petroleum Chemistry, Inc., American Chemical Society, Vol. 32, No. 2, April 1987, pp. 584–590.

[*2*] Wilson, M. F. and Fisher, I. P., *Preprints,* Division of Petroleum Chemistry, Inc., American Chemical Society, Vol. 32, No. 2, April 1987, pp. 572–578.

[*3*] Stephenson, L. C., Muradian, A., Fookes, C. J. R., Atkins, A. R., and Batts, B. D., *Preprints,* Division of Petroleum Chemistry, Inc., American Chemical Society, Vol. 32, No. 2, April 1987, pp. 595–599.

[*4*] Moore, H. C., Johnson, C. A., Sutton, W. A., and Benslay, R. M., *Preprints,* Division of Petroleum Chemistry, Inc., American Chemical Society, Vol. 32, No. 2, April 1987, pp. 578–583.

[*5*] Knudson, C. W., Willson, W. G., Miller, D. J., Ness, R. O., and Ruud, A., *Preprints,* Division of Petroleum Chemistry, Inc., American Chemical Society, Vol. 32, No. 2, April 1987, pp. 591–594.

[*6*] Warner, J. S., Danison, T. H., and McNutly, J. S., "Analytical Techniques for Aromatic Components in Aircraft Fuels," Technical Report AFWAL-TR-82-2015, Air Force Wright Aeronautical Laboratories, OH, 1982.

[*7*] Cooper, J. R. and Taylor, L. T., "GC-FTIR of Jet Fuels," Technical Report AFWAL-TR-84-2007, Air Force Wright Aeronautical Laboratories, OH, 1984.

[*8*] Sink, C. W., Hardy, D. R., and Hazlett, R. N., "Compound Class Quantitation of JP-5 Jet Fuel by High Performance Liquid Chromatography/Differential Refractive Index Detection," NRL Memorandum Report 5407, Naval Research Laboratory, Washington, DC, 1984.

[*9*] Buttrill, S. E., "Analysis of Middle Distillate Fuels by Field Ionization Mass Spectrometry," Interim Technical Report AD-A099407, SRI International, Menlo Park, CA, 1980.' "

[*10*] Hayes, P. C., and Pitzer, E. W., "Kovats Indices as a Tool in Characterizing Hydrocarbon Fuels in Temperature Programmed Glass Capillary Gas Chromatography. Part 1. Qualitative Identification," Technical Report AFWAL-TR-81-2102, Air Force Wright Aeronautical Laboratories, OH, 1980.

[*11*] Rostad, C. E. and Pereira, W. E., *Journal of High Resolution Chromatography & Chromatography Communications,* Vol. 9, June 1986, pp. 328–334.

[*12*] Hawthorne, S. B. and Miller, D. J., *Analytical Chemistry,* Vol. 57, No. 3, March 1985, pp. 694–698.

Jodie V. Johnson,[1] *Richard A. Yost,*[1] *and Carla M. Wong*[2]

Tandem Mass Spectrometry for the Determination of Trace Steranes in Crude and Shale Oils

REFERENCE: Johnson, J. V., Yost, R. A., and Wong, C. M., **"Tandem Mass Spectrometry for the Determination of Trace Steranes in Crude and Shale Oils,"** *Novel Techniques in Fossil Fuel Mass Spectrometry, ASTM STP 1019,* T. R. Ashe and K. V. Wood, Eds., American Society for Testing and Materials, Philadelphia, 1989, pp. 47–58.

ABSTRACT: Triple quadrupole tandem mass spectrometry (MS/MS) has been used to characterize the MS/MS behavior of five sterane standards and then to determine the sterane distribution in two shale oils and a crude oil without prior sample cleanup. Comparisons were made between electron ionization (EI) MS/MS and methane positive chemical ionization (PCI) MS/MS of the sterane standards and the three oils. The EI-MS/MS of the molecular ions ($M^{+\cdot}$) of the steranes was determined to be more sensitive and selective than PCI-MS/MS of the corresponding $(M - H)^+$ ions. With solids probe/EI-MS/MS parent scan of *m/z* 217 (the most characteristic daughter ion for steranes), it was possible to obtain the sterane distribution in the three oils. However, due to chemical interferences, only the most abundant steranes could be confirmed by the complete EI-MS/MS daughter spectra of the detected molecular ions. In addition, severe contamination of the instrument resulted due to introduction of the oils into the vacuum system on the solids probe. Some sample cleanup would be beneficial in order to enhance the ability to detect low-level steranes and to prevent instrument contamination.

KEY WORDS: tandem mass spectrometry, steranes, shale oil, crude oil, triple quadruple tandem mass spectrometry, mass spectrometry, fossil fuels

Steranes are a class of biomarkers used in fossil fuel exploration to ascertain the origin (terrestrial or marine), maturity, migration, and chemical and biodegradation changes of the fossil fuels [*1–3*]. Due to low levels (ppm) of individual steranes in oil and the complexity of the sterane composition, extensive cleanup procedures and gas chromatography (GC) temperature programming have been required, resulting in lengthy analyses with the traditional GC/EI-MS (electron ionization-mass spectrometry) methods of analysis [*1–7*]. More recently, tandem mass spectrometry (MS/MS) has been shown to significantly enhance the selectivity for the determination of steranes and other biomarkers in oil [*1,8–12*]. A recent book and several review articles describe the instrumentation, methods, and various applications of MS/MS [*13–16*]. Here, we report the results of our evaluation of the use of triple quadrupole tandem mass spectrometry to characterize the MS/MS behavior of steranes and to rapidly determine the sterane distribution in two shale oils and a crude oil without prior sample cleanup.

[1] Assistant research scientist and associate professor, respectively, Chemistry Department, University of Florida, Gainesville, FL 32611.

[2] Chemist, Lawrence Livermore National Laboratory, Livermore, CA 94550; currently at NASA, Ames Research Center, Moffett Field, CA 94035.

Experimental

Samples

The sterane standards (Table 1) and oil samples were provided by Lawrence Livermore National Laboratory (LLNL) (Livermore, California). The standards were dissolved in benzene to obtain low parts-per-thousand to high parts-per-million standard solutions. Each of the oil samples was also dissolved in 100 μL of benzene. The oil samples consisted of a Fischer assay oil (G-3 oil) prepared from shale from the Green River Formation at Logan Wash, a seep oil (TC-15 oil) from the White River, Utah, oil shale mine, and a Fischer assay oil (Eastern Oil-A) from an eastern shale (New Albany Shale from Bullet County, Kentucky).

Instrumentation

All data were collected with a Finnigan MAT (San Jose, California) TSQ45 triple quadrupole tandem mass spectrometer equipped with a 4500 series ion source, pulsed positive ion and negative ion chemical ionization (PPINICI), and INCOS data system. A Finnigan 9610 gas chromatograph, equipped with a Grob-type capillary column injector, was interfaced to the TSQ45.

Sample Introduction

Gas Chromatography—For sample introduction via gas chromatography (GC), samples were injected onto a 10 m, 0.25 mm inside diameter DB-5 bonded-phase fused silica capillary GC column (J & W Scientific, Folsom, California) inserted directly into the ion source. Sterane standards were injected either in the splitless mode (split and sweep closed during and after injection for 1 min) or in the split mode (split and sweep open at all times) with a split flow rate of 50 mL/min and a septum sweep flow rate of 5 mL/min and with an average helium carrier gas velocity of approximately 65 cm/s (measured by the elution of *n*-hexane at 250°C). For splitless injections, the GC oven was held at 50°C for 1 min following injection and, thereafter, increased linearly at a rate of 20°C/min to a maximum of 275°C. For split injections, the GC oven was held isothermally at a temperature to allow fast elution of the steranes but separation from the benzene solvent. The interface and injection port temperatures were maintained at 275°C.

For analysis of the oil samples, the injection port temperature was increased to 300°C while maintaining all the other temperature regimes the same as before. Injections of 2.1 μL of the oil solutions (the viscous residues of the oil samples were dissolved in 100 μL benzene) were made in the splitless mode (1 min) and with the same GC temperature

TABLE 1—*Molecular formulas and weights for the sterane standards examined.*

Sterane	Molecular Formula	Monoisotopic Molecular Weight
5-α-androstane	$C_{19}H_{32}$	260.2
5-α-pregnane	$C_{21}H_{36}$	288.3
5-α-cholestane	$C_{27}H_{48}$	372.4
5-α-ergostane	$C_{28}H_{50}$	386.4
5-α-stigmastane	$C_{29}H_{52}$	400.4

program. The split and sweep flow rates were 10 and 2 mL/min, respectively, and the average linear velocity was approximately 100 cm/s as measured by the elution of *n*-hexane at 300°C.

Solids Probe—For introduction of the oil samples via the direct insertion solids probe, 5 μL of an oil solution was "dried" in a glass capillary vial at 60°C for ~1 h while exposed to the room air. The glass vial was then inserted directly into the ion source via the direct insertion solids probe. In the initial study of the oils, vaporization of the sample was achieved by heating the solids probe under computer control from approximately 50 to 400°C at 100°C/min. However, to reduce the contamination of the ion source in later experiments, the solids probe was introduced into the vacuum system but left outside of the ion source. The solids probe was heated to and held at 100°C for 1 min to allow the lower boiling materials to vaporize. The probe was then inserted fully into the ion source and heated to 200°C at 50°C/min followed by heating to 400°C at 100°C/min.

Mass Spectrometry

Normal Mass Spectrometry—The ion optics were tuned and the mass range calibrated for the Q1 and Q3 normal mass spectral modes with the ions generated from the electron ionization (EI, 70 eV electron energy, 0.3 mA emission current) of perfluorotri-*n*-butylamine (PFTBA or FC43). The ion optics were tuned such that the *m/z* 69 ion was the most abundant ion and the following ion intensity ratios were obtained: 131/69 = 0.4 to 0.7, 219/69 = 0.5 to 0.7, and 502/220 = 1 to 3. For chemical ionization (CI, 100 eV electron energy, indicated source pressure of 0.9 to 1.0 torr methane, ion source temperature of 150 °C), the ion optics were tuned to maximize the intensities of the *m/z* 414 positive and negative ions of PFTBA in both the positive CI (PCI) and electron-capture negative CI (ECNCI) modes. Standards were introduced on the capillary column and the EI, ECNCI, and PCI normal mass spectra were obtained in the Q3 normal mass spectrum mode. For GC/MS analysis of the oil samples, scanning over a wide mass range (35 to 460 atomic mass unit in 1 s) was not sensitive enough to adequately detect the presence of any steranes. Therefore, selected ion monitoring (SIM) of the EI-generated *m/z* 217 and the molecular ions of the steranes between *m/z* 260 and 470 were monitored in the Q1 normal mass spectrum mode with a 1 atomic mass unit wide window centered at each *m/z*.

Tandem Mass Spectrometry (MS/MS)—For MS/MS, the instrument was tuned separately for EI-CAD and PCI-CAD and for parent and daughter scan modes. As the ion transmission of some of the ion optics are mass-dependent, this permitted optimal transmission of the ions of interest, that is, parent ions or daughter ions. In the daughter scan mode, the ion optics were tuned to maximize the intensities of the *m/z* 169 daughter ions of the EI- and PCI-generated *m/z* 414 ions of PFTBA at a collision gas pressure and collision energy used for the subsequent sample analyses. For MS/MS in the parent scan mode, the ion optics were tuned to maximize the intensity of the *m/z* 414 and 502 parent ions of *m/z* 169 and 219, respectively, of PFTBA with both EI and PCI and at the CAD conditions used for the subsequent sample analyses. For analysis of standards, collision gas pressures of 1.1 and 1.75 mtorr nitrogen and collision energies of 5, 15, and 25 eV were used. The analyses of the oil samples were performed with a collision gas pressure of 1.1 mtorr nitrogen and a collision energy of approximately 25 eV. The daughter spectra that are reported are the average of several of the most intense spectra obtained during the elution of the steranes from the GC or off of the solids probe. The parent spectra shown for the solids probe/MS/MS of the oil samples are the integration of all the spectra acquired during the vaporization of the sample from the solids probe.

Results and Discussion

For the analysis of mixtures by MS/MS, it is desirable that the initial ionization of a mixture yields a minimum number of ions (ideally, one) per component and that these ions are characteristic of the parent compounds. Traditionally, chemical ionization has been used to achieve this goal. However, it is also desirable that the MS/MS of the ionized compounds of interest yield characteristic daughter ions. Without *a priori* knowledge, therefore, it is wise to characterize the MS and MS/MS behavior of the compounds of interest for the ionizations that are available. Here we have compared the EI-MS/MS and methane CI-MS/MS for the determination of steranes in oils.

Normal Mass Spectrometry

The EI and methane CI normal mass spectra of each of the sterane standards were acquired to evaluate the characteristic ions resulting from the two different ionizations. The EI mass spectra of the sterane standards were dominated by fragment ions, most notably *m/z* 217 (illustrated by 5-α-stigmastane in Fig. 1*a*). Although each of the steranes yielded a molecular ion, its relative ion intensity decreased with the increasing length of the alkyl-chain substituent to the five-member ring.

Although methane-enhanced ECNCI was attempted, as expected, no response was observed for any of the compounds of interest. This was due to the lack of any electronegative atoms in the molecules, making ECNCI transparent to these hydrocarbons.

The methane PCI yielded an abundant $(M - H)^+$ ion for each of the steranes (Fig. 1*b*). No $(M + H)^+$ or adduct ions were observed for any of the steranes. Although the steranes

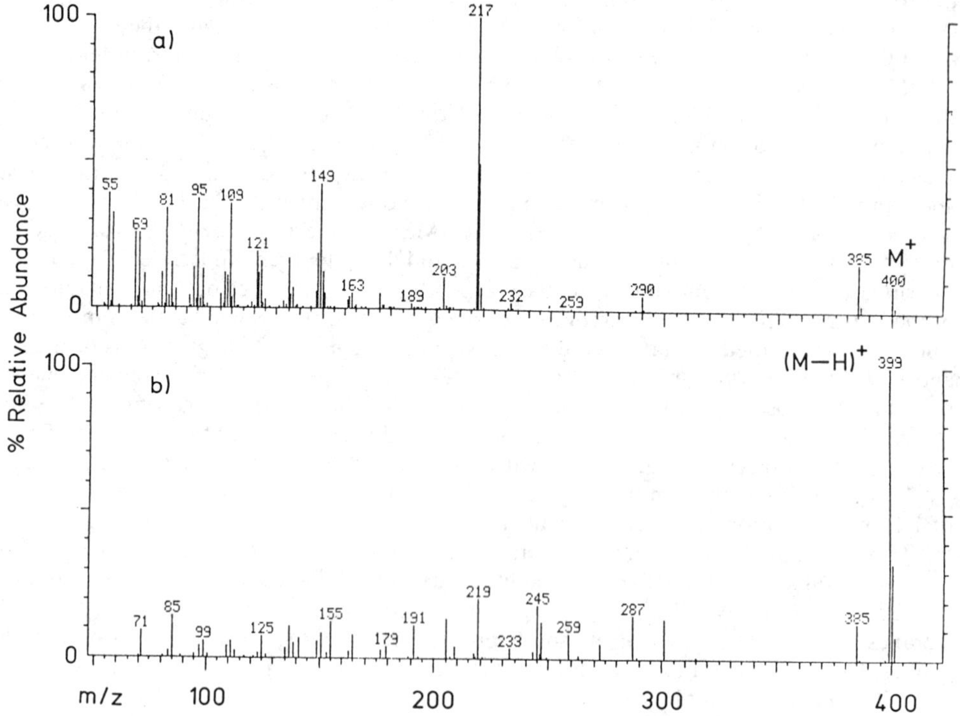

FIG. 1—*The* (a) *EI and* (b) *methane PCI normal mass spectra for 5-α-stigmastane.*

all underwent extensive fragmentation as evidenced by the presence of numerous fragment ions, none of the fragment ions exceeded 20% relative abundance. Rather, the spectra are dominated by the $(M - H)^+$ ions. Thus, methane PCI, lacking the extensive fragmentation that occurs in EI, would be expected to offer a little more sensitivity and selectivity than EI would. The sensitivities for the EI $M^{+\cdot}$ and the PCI $(M - H)^+$ sterane ions were compared by GC/selected ion monitoring (SIM) of each of these ions. For all of the steranes examined, the PCI-SIM of the $(M - H)^+$ ions was 2 to 12 times more sensitive than EI-SIM, with the degree of enhancement decreasing with the increasing molecular weight of the sterane. As PCI had the greater sensitivity and less fragmentation, it was initially thought to be the preferred ionization for analyzing the oil samples. However, as will be discussed later, the characteristics of the daughter mass spectra must also be taken into account.

Tandem Mass Spectrometry

Initially, the collisionally activated dissociation (CAD) daughter mass spectra of the EI $M^{+\cdot}$ and the PCI $(M - H)^+$ ions were obtained with a collision gas pressure of 1.75 mtorr nitrogen and collision energies of 22 and 25 eV, respectively. However, these conditions were rather harsh, producing too much fragmentation. Subsequently, the collision gas pressure was lowered to approximately 1.1 mtorr nitrogen and the CAD daughter spectra were acquired at three different collision energies (5, 15, and 25 eV) in a single GC run of each standard. Although some interesting structural information was seen in the daughter spectra acquired at the lower collision energies, the CAD efficiencies and MS/MS sensitivities were highest at 25 eV.

The EI-CAD Daughter Mass Spectra—The EI-CAD (1.1 mtorr N_2, 25 eV) daughter mass spectra of the $M^{+\cdot}$ ions of the sterane standards are characterized by abundant $(M - 15)^+$ and 217^+ daughter ions, with the daughter mass spectra becoming dominated by these two daughter ions with increasing molecular weight of the sterane (Table 2 and shown in Fig.

TABLE 2—*The EI-CAD (1.1 mtorr nitrogen, 25 eV) daughter spectra of the molecular ions ($M^{+\cdot}$) of the steranes.*

			Daughter Ions					
Sterane	$M^{+\cdot}$		$(P - 110)^+$	217^+				
5-α-androstane	260(44)	→	150(7)[a]	217(25)	245(100)	218(11)	204(21)	203(53)
					189(16)	161(28)	148(24)	135(13)
5-α-pregnane	288(15)	→	178(14)	217(100)	273(60)	218(52)	203(10)	175(11)
					161(6)	149(6)	148(12)	147(6)
5-α-cholestane	372(9)	→	262(9)	217(100)	357(59)	218(8)	203(3)	162(4)
					148(4)	147(4)	122(3)	121(5)
5-α-ergostane[b]	386(ND)[c]	→	276(<10%)	217(100)	274(10)	147(12)	135(18)	
5-α-stigmastane	400(7)	→	290(6)	217(100)	385(57)	287(2)	218(3)	162(4)
					148(4)	147(6)	122(8)	121(3)

[a] Percent relative abundance (%RA) normalized to the most abundant daughter ion in the daughter spectrum. In addition to the $(P - 110)^+$ and 217^+ daughter ions, only the eight most intense daughter ions with %RA > 1% are shown.

[b] The EI-CAD daughter spectrum of 5-α-ergostane was obtained at 1.7 mtorr nitrogen and 25 eV. Due to the small amount of sample and the poor signal-to-noise ratio, only ions greater than 10% RA are reported for this sterane.

[c] ND = not detected (<0.5% RA).

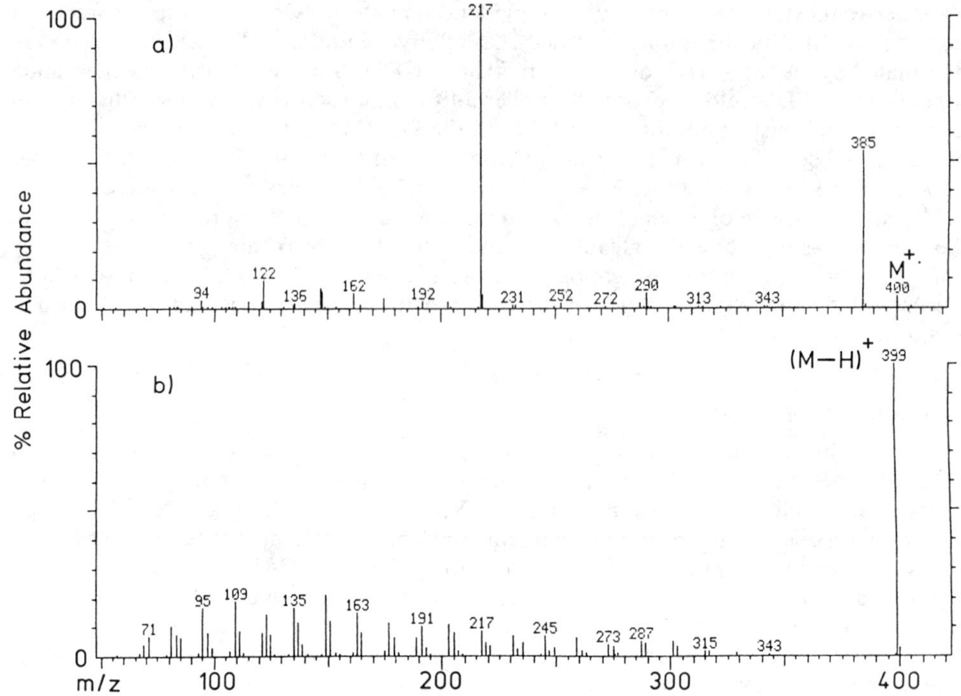

FIG. 2—*The CAD (1.1 mtorr nitrogen and 25 eV) daughter spectra of the* (a) *EI* $M^{+\cdot}$ *ion and* (b) *the methane PCI* $(M - H)^+$ ion of 5-α-*stigmastane.*

2*a* for 5-α-stigmastane). This same behavior was observed in the normal EI mass spectra just discussed. As the *m/z* 217 daughter ion was an abundant ion in all of the daughter spectra, it served as the common daughter ion in the subsequent MS/MS parent scan screening of oil samples for steranes. In addition, the loss of a neutral fragment of 110 atomic mass unit from the parent molecular ion yielded a characteristic daughter ion, for example, *m/z* 290 for the 5-α-stigmastane.

The PCI-CAD Daughter Mass Spectra—Unlike the EI-CAD of the M^+, the methane PCI-CAD (25 eV, 1.1 mtorr N_2) daugher spectra of the $(M - H)^+$ ions were not dominated by one or two major daughter ions (Table 3 and Fig. 2*b*). Rather, they were characterized by numerous daughter ions that comprised four major ion series: 91 + $(CH_2)_n$, 79 + $(CH_2)_n$, 67 + $(CH_2)_n$, and 69 + $(CH_2)_n$, with the most characteristic and usually the most abundant daughter ion in each series being *m/z* 203, 149, 95, and 111, respectively. As these ions were the most dominant and seem to be fairly characteristic, they were chosen to be the selected daughter ions in the PCI-CAD parent screening of the oil samples below. It should also be noted that, although methane PCI yields a much greater relative abundance of the $(M - H)^+$ ions than does EI for the $M^{+\cdot}$ ions, the $(M - H)^+$ ions are much more stable towards CAD and are not as efficiently fragmented. Thus, while PCI-MS may be more sensitive than EI-MS, PCI-MS/MS was expected to be less sensitive than EI-MS/MS.

The MS/MS Analysis of Oil Samples

Gas Chromatography/Mass Spectrometry—Initially, the sterane-rich TC-15 oil was analyzed by traditional GC/EI-MS with a 10-m DB-5 fused silica capillary GC column. How-

TABLE 3—*The PCI-CAD (1.1 mtorr nitrogen and 25 eV) daughter spectra of the $(M - H)^+$ ions of the sterane standards.*

Sterane	$(M - H)^+$		Daughter Ions 203^+					
5-α-androstane	259(148)	→	203(14)[a]	177(20)	163(73)	149(100)	137(11)	135(54)
				123(22)	121(23)	109(42)	95(43)	81(29)
5-α-pregnane	287(333)	→	203(26)	191(66)	177(92)	163(97)	149(94)	137(58)
				135(63)	123(100)	109(73)	95(72)	81(52)
5-α-cholestane	371(551)	→	203(49)	177(49)	163(78)	151(54)	149(100)	135(75)
				123(56)	111(49)	109(89)	95(96)	81(50)
5-α-ergostane	385(468)	→	203(64)	191(49)	151(49)	149(100)	137(45)	135(75)
				123(67)	111(52)	109(67)	95(98)	81(62)
5-α-stigmastane	399(483)	→	203(52)	191(49)	177(56)	163(72)	151(58)	149(100)
				137(54)	135(80)	123(68)	109(90)	95(78)

[a] Percent relative abundance (%RA) normalized to the most abundant daughter ion in the daughter spectrum. In addition to the 203^+ daughter ion, only the ten most abundant daughter ions with %RA > 1% are shown.

ever, after only five oil analyses, the sensitivity decreased by a factor of 10 due to the deterioration of column performance by severe contamination from the oils. Due to this column deterioration and to the limited amount of oil solution that could be injected, the direct insertion solids probe was investigated as a means of sample introduction.

Solids Probe MS/MS—Solids probe (SP) MS/MS was evaluated for the determination of steranes in the shale and crude oils by the EI-CAD parent scan of 217^+ and the PCI-CAD parent scans of 95^+, 149^+, and 203^+. Figure 3 illustrates the results of such screenings for the sterane-rich TC-15 oil. The EI-CAD parent spectra of m/z 217 were dominated by an even m/z ion series (m/z 288, 302, 316, 330, 344, 358, 372, 386, 400, and 414) characteristic of sterane molecular ions (Fig. 3*a*). The PCI-CAD parent spectra of m/z 95, 149, and 203 contained the corresponding ion series indicative of the $(M - H)^+$ ions of the steranes (Figs. 3*b*, 3*c*, and 3*d*, respectively). As the m/z of the selected daughter ion increased, the selectivity of the PCI-CAD parent scan also increased, with the parent scan of m/z 203 yielding the greatest selectivity of the three. However, even in the PCI-CAD parent spectrum of m/z 203, note the presence of many other non-sterane ions that are largely lacking in the EI-CAD parent spectrum of m/z 217. In addition, the parent scan of m/z 203 had the lowest sensitivity of the three PCI-CAD parent scans, and all of the PCI-CAD parent scans were less sensitive than the EI-CAD parent scan of m/z 217. As mentioned earlier, this was due to the EI $M^{+\cdot}$ ions fragmenting to yield largely a single daughter ion, m/z 217, whereas the PCI $(M - H)^+$ ions fragmented inefficiently to form a large number of more equally abundant, but less intense, daughter ions. Thus, for increased MS/MS selectivity and sensitivity, the EI-CAD parent scan of m/z 217 would be the preferred method for screening for steranes.

The EI-CAD parent spectra of m/z 217 of each of the three oils examined were significantly different (Fig. 4). If the intensities of each of the molecular ions is normalized to the intensity of the m/z 217 daughter ion and the assumption is made that the composition of the detected sterane-type molecular ions is indeed largely steranes, some semi-quantitative characterization of the oils can be obtained (Table 4). Normalized as such, the TC-15 oil had the most abundant sterane content, followed by the G-3 oil, and then the Eastern Oil-A. Whereas the Eastern Oil-A spectrum contained largely ions indicative of the low m/z

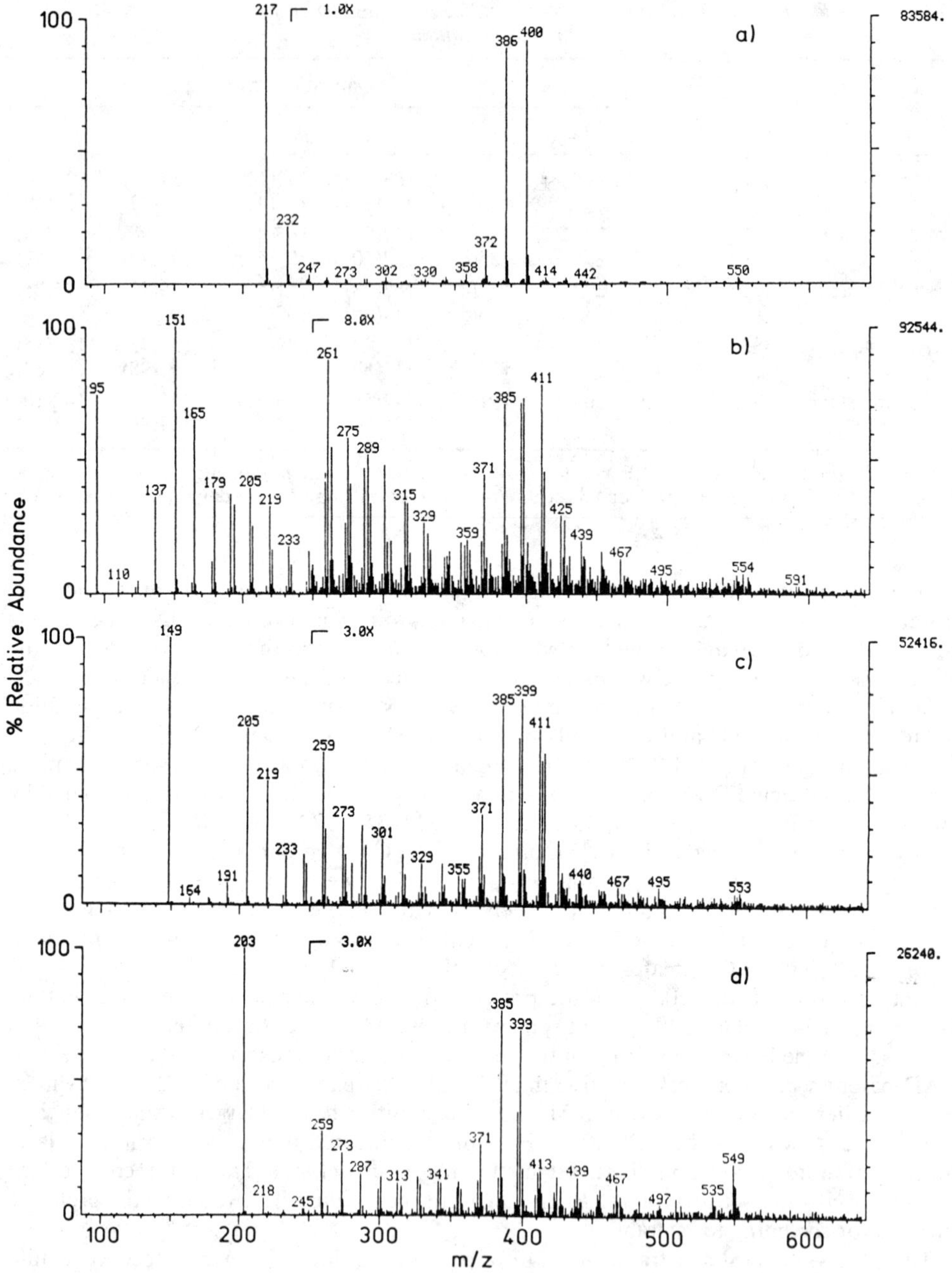

FIG. 3—*The CAD (1.1 mtorr nitrogen and 25 eV) parent spectra of the sterane-rich TC-15 oil:* (a) *EI-CAD parents of* m/z *217,* (b) *methane PCI-CAD parents of* m/z *95,* (c) *methane PCI-CAD parents of* m/z *149, and* (d) *methane PCI-CAD parents of* m/z *203.*

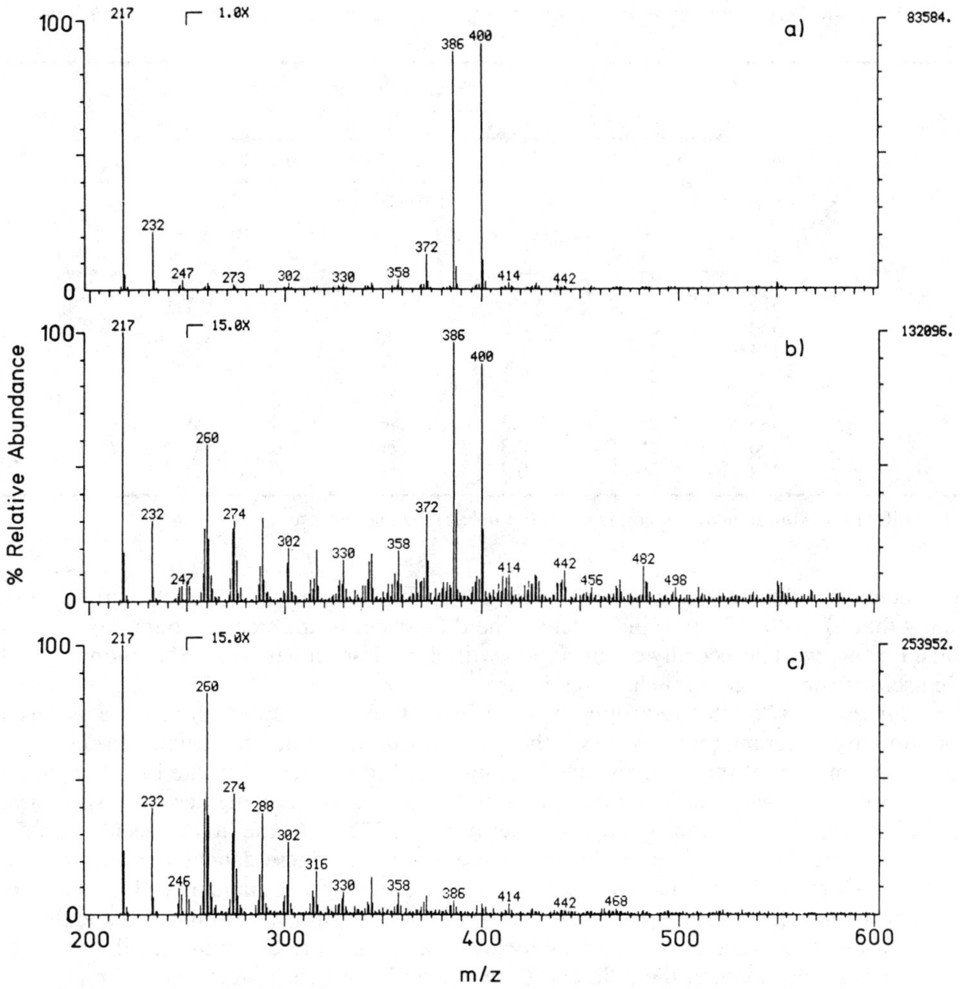

FIG. 4—*The EI-CAD parent spectra of* m/z *217 obtained for three different oils:* (a) *TC-15 oil,* (b) *G-3 oil, and* (c) *Eastern Oil-A.*

sterane molecular ions (that is, *m/z* 260 for a C_{19} sterane) and decreasing abundance of steranes with increasing carbon numbers, both the TC-15 and G-3 oils showed a maximum sterane content for the C_{28} and C_{29} steranes at *m/z* 386 and 400, respectively. The sterane content of the TC-15 and G-3 oils agrees with the general finding that the C_{27}, C_{28}, and C_{29} steranes are the most abundant steranes in oils [*1,6*]. It is also interesting to note that there are sterane-type molecular ions detected above *m/z* 400 (that is, 414, 428, 442, etc.). Generally, the steranes detected by traditional GC/MS are limited to those below C_{30} due to the upper temperature limits (350°C) of most GC columns and the volatility of the steranes. With the use of SP/MS/MS, these restrictions are much less severe, offering the capability to study higher carbon number steranes. Similar results have been obtained in the study of the hopane distributions where MS/MS extended the range of hopanes thought to exist from C_{40} to C_{70} [*1*].

It would be interesting to correlate these observed differences in sterane content to the

TABLE 4—*Relative abundances of steranes in three different oils as determined by the solids probe EI-CAD parent scan of* m/z *217.*

		% Relative Abundance[a]		
m/z	Sterane, Carbon Number	Eastern Oil A	Oil TC-15	Oil G-3
260	19	6	2	4
274	20	3	1	2
288	21	2	2	2
302	22	2	2	1
316	23	1	1	1
330	24	0.5	2	1
344	25	1	2	1
358	26	0.5	3	1
372	27	0.5	13	2
386	28	0.3	88	6
400	29	0.3	91	6
414	30	0.3	2	0.7

[a] % Relative Abundance is normalized to the *m/z* 217 daughter ion.

geochemical differences in the oils. However, these are preliminary results and serve to show that SP/MS/MS can rapidly determine differences in the sterane content of several oils. Further work, especially confirmation with more traditional GC/MS methods, would be useful prior to making such correlations.

Solids Probe MS/MS Confirmation—The EI-CAD parent scan of *m/z* 217 served as a preliminary screening tool to assess if there is a possibility of steranes being present in the oil. Confirmation of the steranes that were indicated to be present by the EI-CAD parent scan of *m/z* 217 was attempted by obtaining full scan EI-CAD daughter mass spectra of the detected molecular ions (shown for the *m/z* 358, 372, 386, and 400 molecular ions of the C_{26}, C_{27}, C_{28}, and C_{29} steranes, respectively, in Fig. 5). These daughter spectra do not agree very well with the daughter spectra acquired for the sterane standards with the exception of the daughter spectra of the *m/z* 386 and 400 ions observed in the G-3 and TC-15 oils. Even in these spectra, there are many ions that are not present in the daughter spectra of the standards. Some of the differences are due to the fact each molecular ion represents many different sterane positional isomers and diastereomers and not a single sterane. However, the presence of *m/z* 217 as the most abundant daughter ion and the presence of the readily observable $(P - 110)^+$ daughter ions (that is, the *m/z* 276 daughter ion of the *m/z* 386 parent ion in Fig. 5*c* and the *m/z* 290 daughter ion of the *m/z* 400 parent ion in Fig. 5*d*) are good evidence that the *m/z* 386 and 400 molecular ions indeed correspond to sterane-type molecules (Figs. 5*c* and *d*). It was not possible to confirm the presence of any of the other possible steranes by the full daughter spectra due to the abundant chemical interferences (note the quality of the *m/z* 358 and 372 daughter spectra of Figs. 5*a* and *b*, respectively). Confirmation of the presence of these less abundant steranes would require some sample clean-up and concentration.

Contamination and Chemical Interferences—Although the EI-CAD parent scan of *m/z* 217 was successful at rapidly screening a small number of oil samples for steranes, this is not a good method for analyzing large numbers of oil samples without some type of chemical cleanup. The oils never really dried and upon heating in the vacuum system, severe contamination of the instrument resulted in a rapid deterioration in instrument performance; after only 12 analyses, the entire ion source and analyzer region had to be disas-

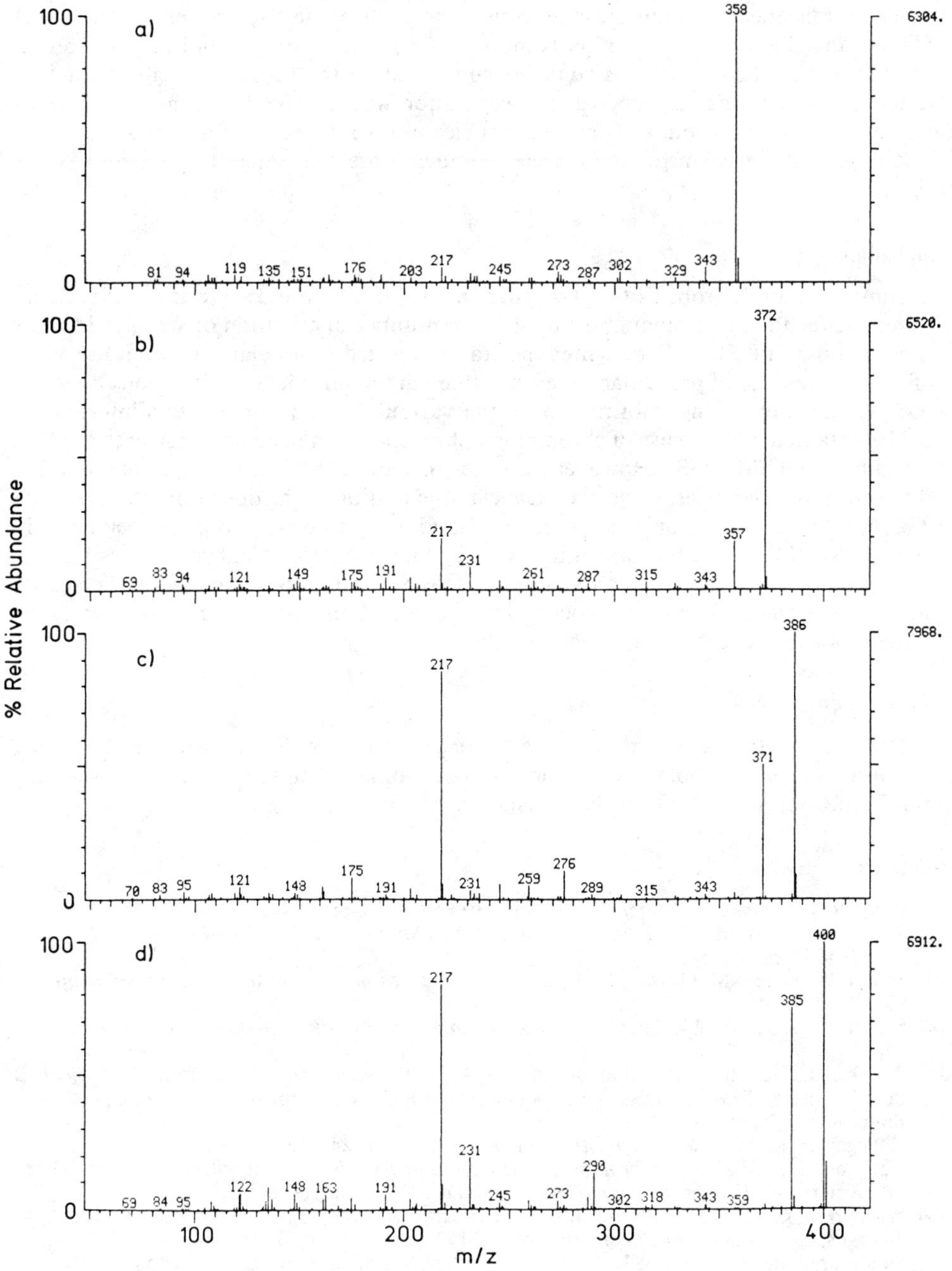

FIG. 5—*The EI-CAD daughter spectra of* (a) m/z *358,* (b) m/z *372,* (c) m/z *386, and* (d) m/z *400 obtained for the TC-15 oil.*

sembled and cleaned. In addition, due to the low levels of steranes present in shale and crude oils and the multitude of other components present creating chemical interference, only the most abundant steranes could be confirmed by the EI-CAD daughter scans of detected molecular ions. A chemical class separation would significantly improve both of these concerns. In addition, such a chemical cleanup would also enable the use of short GC columns for sample introduction that are much more convenient to use than a solids probe.

Conclusions

In summary, solids probe EI-CAD parent scan of m/z 217 would serve as a rapid screening method for the determination of the carbon number distribution of steranes in shale and crude oils. The EI-CAD daughter spectra of detected molecular ions would serve to confirm the presence of particular steranes. Although the introduction of the oils by solids probe precludes obtaining information on the various sterane isomers, this information could be obtained with the use of chromatographic separation in conjunction with MS/MS (for example, GC/MS/MS). However, in order to reduce the instrument contamination and chemical interferences, some chemical class separation of the oils prior to SP/MS/MS or GC/MS/MS analysis would be necessary. In addition to being a rapid screening technique, SP/MS/MS offers the capability to study higher carbon number steranes (greater than C_{30}) than can be done by conventional GC/MS or GC/MS/MS methods. The correlation of these steranes with the geochemical records of oils may serve as another useful parameter for the geochemist in studying oils.

Acknowledgment

We would like to thank Arthur Lewis, manager of the Oil Shale Project at Lawrence Livermore National Laboratory, for providing funding for this research. We especially thank Tom Coburn of LLNL for his assistance.

References

[*1*] Philp, R. P., *Chemical & Engineering News,* 10 Feb. 1986, pp. 28–43.
[*2*] Seifert, W. K., Moldowan, J. M., and Jones, R. W., *Spectra,* Finnigan MAT Corp, San Jose, CA, Vol. 8, 1982, pp. 42–51.
[*3*] Seifert, W. K. and Moldowan, J. M., *Geochimica et Cosmochimica Acta,* Vol. 42, 1978, pp. 77–95.
[*4*] Seifert, W. K. and Moldowan, J. M., *Geochimica et Cosmochimica Acta,* Vol. 43, 1979, pp. 111–126.
[*5*] Petrakis, L., Jewell, D. M., and Benusa, W. F. in *Petroleum in the Marine Environment,* L. Petrakis and F. T. Weiss, Eds., American Chemical Society, Washington, DC, 1980, pp. 23–53.
[*6*] Philp, R. P., *Spectra,* Vol. 8, 1982, pp. 6–31.
[*7*] Gallegos, E. J., *Analytical Chemistry,* Vol. 47, 1975, pp. 1524–1528.
[*8*] Gallegos, E. J. in *High Performance Mass Spectrometry: Chemical Applications,* M. L. Gross, Ed., American Chemical Society, Washington, DC, 1978, Chapter 15.
[*9*] Warburton, G. A. and Zumberge, J. E., *Analytical Chemistry,* Vol. 55, 1983, pp. 123–126.
[*10*] Gallegos, E. J., *Analytical Chemistry,* Vol. 48, 1976, pp. 1348–1351.
[*11*] Sundararaman, P., Gallegos, E. J., Baker, E. W., Slayback, J. R. B., and Johnston, M. R., *Analytical Chemistry,* Vol. 56, 1984, pp. 2552–2556.
[*12*] Johnson, J. V., Britton, E. D., Yost, R. A., Quirke, J. M. E., and Cuesta, L. L., *Analytical Chemistry,* Vol. 58, 1986, pp. 1325–1329.
[*13*] *Tandem Mass Spectrometry,* F. W. McLafferty, Ed., Wiley, New York, 1983.
[*14*] Cooks, R. G. and Glish, G. L., *Chemical & Engineering News,* 30 Nov. 1981, pp. 40–52.
[*15*] Johnson, J. V. and Yost, R. A., *Analytical Chemistry,* Vol. 57, 1985, pp. 758A–768A.
[*16*] Yost, R. A. and Fetterolf, D. D., *Mass Spectrometry Reviews,* Vol. 2, 1983, pp. 1–45.

Gareth E. Harriman,[1] Raymond Owen,[2] Victor C. Parr,[2] Orla Weir,[2] and David Wood[2]

A Comparison of the Methods of Medium Resolution Selected Ion Recording and Multiple Reaction Monitoring When Applied to the Analysis of Steranes and Triterpenes in Crude Oil Extracts

REFERENCE: Harriman, G. E., Owen, R., Parr, V. C., Weir, O., and Wood, D., **A Comparison of the Methods of Medium Resolution Selected Ion Recording and Multiple Reaction Monitoring When Applied to the Analysis of Steranes and Triterpenes in Crude Oil Extracts,"** *Novel Techniques in Fossil Fuel Mass Spectrometry, ASTM STP 1019,* T. R. Ashe and K. V. Wood, Eds., American Society for Testing and Materials, Philadelphia, 1989, pp. 59–83.

ABSTRACT: Oil correlation studies and assessments of the environment of deposition of source rocks and of the maturation, migration, and biodegradation of crude oils are commonly based upon the distribution of "biological marker" compounds within the oil or source rock extract. The measurement of these compounds involves the use of combined capillary column gas chromatography with mass spectrometry. However, the compounds are usually present at low levels and so selected ion recording techniques are preferred. Comparisons are made of two methods for increasing the selectivity of the mass spectrometer beyond that available when using nominal mass resolution, such as that available with quadrupole instruments. These methods are, increasing the mass spectrometer resolving power to, say 2 to 3000, and secondly to monitor fragments formed in the first field free region of a double-focusing instrument. The second method generally yields data that are easier to interpret and are less confused by interferences. Moreover, the sensitivity is adequate to obtain results using whole oils and total source rock extracts without prior sample cleanup and concentration.

KEYWORDS: biomarkers, steranes, triterpenes, gas chromatography-mass spectrometry, multiple reaction monitoring, metastable ions, oil-oil correlation, fossil fuels, mass spectrometry

The study of biological marker (biomarkers) compounds in crude oils and source rock extracts is now an established method used by petroleum explorationists in identifying genetic relationships between various oils and in aiding oil-to-oil and oil-to-source rock correlations.

Polycyclic hydrocarbons such as steranes and triterpenes have been found to be the most useful biomarkers for such studies and their use for the assessment of environment of

[1] Senior geochemist, Geochem Laboratories, Chester, UK.

[2] Design engineer, managing director, applications chemist, and technical director, respectively, VG Tritech, Wythenshawe, Manchester, UK.

deposition, maturation, migration, and alteration of crude oils is widely documented [*1–4*].

The "typing" of crude oils and source rock extracts is normally carried out by gas chromatography-mass spectrometry (GC-MS), usually with a quadrupole (or similar) mass spectrometer operated in a low (unit mass) resolution mode. Unfortunately, when monitoring in selected ion recording mode at low resolution, responses at m/z 191 and 217 (base fragment ions for triterpenes and steranes, respectively) are detected not only from triterpenes and steranes but also from other components within the oil or rock extract. In cases where the concentration of biomarkers is particularly low, these chemical background signals become a significant problem leading to spurious or co-eluting peaks that hinder both identification and quantification of compounds of interest. These problems have, in the past, been overcome by pre-concentrating or isolating the biomarker components prior to GC-MS. This, however, involves complicated and lengthy chemical and chromatographic separation techniques and can often produce misleading results.

Recent advances in mass spectrometry [*5–9*] in geochemistry have shown that crude oils can be analyzed, with little or no sample fractionation required, by operating the mass spectrometer at higher (resolving power of 2 to 3000) resolution. Because this technique is highly specific, when coupled with high-resolution gas capillary columns the analysis allows rapid and quantitative assessment of biomarkers within crude oils without fractionation. One drawback, however, is that the specificity improvement may not be sufficient, especially when the fragment ions are monitored. In some cases, the interfering peaks may be from ions with the same elemental composition. Another approach is to monitor specific decompositions taking place in the first field free region of a double-focusing mass spectrometer [*6,9*]. This technique may offer greater selectivity.

This paper reports on the comparisons of the three methods mentioned for the detection of steranes and triterpenes in two quite different oils, one of marine origin and the second of mature terrestrial origin in which the concentration of biomarkers is very low. Two sterane standards are also used to illustrate the relative detection limits for the three techniques and the nature of the spectra obtained.

Procedure

Equipment

All analyses are carried out using a VG TS-250, double-focusing mass spectrometer to which an HP5890 gas chromatograph (GC) is directly coupled. Samples are injected using a split ratio of 25:1 onto a 25-m bonded-phase, OV-1 fused-silica capillary column. The capillary column passes directly into the ion source via a heated interface line. The program conditions for the GC are shown in Fig. 1.

Table 1 lists the masses monitored during the analysis, the total cycle times were, 1.0 s for the multiple reaction monitoring (MRM) experiment, 1.2 s for low-resolution selected ion recording (SIR), and 1.3 s for the medium-resolution SIR. The latter is longer due to the addition of a lock mass at m/z 207.033 from column bleed. The lock mass feature compensates for any drift in the system.

The mass spectrometrical methods used benefit from two particular features of the instrument chosen for this work. First, the system uses a nonferrous core electromagnet that allows very rapid switching between masses (between 10 and 30 ms) by magnetic switching, even when operating at elevated resolving powers. More conventional iron core systems are limited both in speed and precision by hysteresis and eddy-current effects [*10*]. On such systems, switching must be carried out by changing the accelerating voltage thus

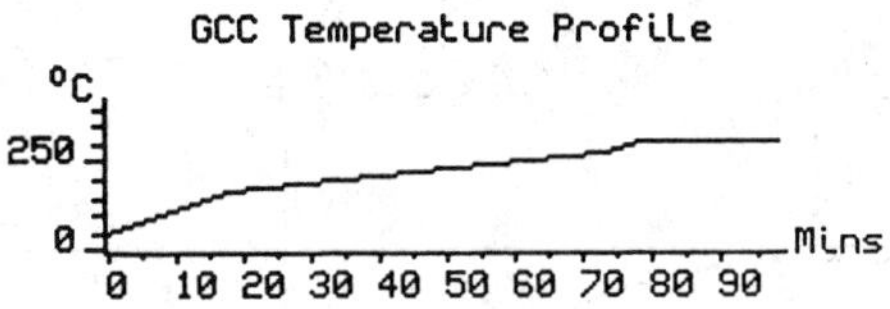

VAP Valve Program	1
VAT Valve Time	.50
INJ Injector Temp	300.0
GCI GC Type	HP589R

TE1 Temperature 1(deg C)	60
TI1 Time 1(mins)	0.00
RT1 Rate 1(deg C/min)	6.0
TE2 Temperature 2(deg C)	170
TI2 Time 2(mins)	0.00
RT2 Rate 2(deg C/min)	2.0
TE3 Temperature 3(deg C)	280
TI3 Time 3(mins)	0.00
RT3 Rate 3(deg C/min)	6.0
TE4 Temperature 4(deg C)	310
TI4 Time 4(mins)	20.00
RT4 Rate 4(deg C/min)	0.0
TE5 Temperature 5(deg C)	0
TI5 Time 5(mins)	0.00

FIG. 1—*Gas chromatography conditions used for the analysis.*

limiting the mass range. There are no such mass range limitations with the chosen instrument. The second point is illustrated in Fig. 2, the system uses a "tri-sector" geometry that incorporates a second electric sector after the electromagnet. A set of conditions are illustrated for one of the transitions monitored in the experiment, namely, the *m/z* 414–217 case. One of the problems associated with monitoring unimolecular dissociations is that they will continue to take place throughout the entire beam path, not just in the first field free region. In the schematic, such a reaction is shown taking place between the magnet and the second electric sector; clearly these new daughter ions will not be transmitted to the final detector in this system, whereas in a more conventional double-sector system they would be seen at the final detector. The more likely possibility is that a completely different species may produce ions between the analyzers that may be observed in a conventional system but will be deflected by the second electric sector, thus reducing the occurrence of "artifacts."

Sample Preparations

The two oils analyzed are crude oils as they are obtained from the well head. No sample cleanup is carried out prior to analysis. The two standards are made up in hexane solution prior to injection.

TABLE 1—*List of masses monitored for the analysis.*[a]

Mass, Amu[b]		Mass, Amu	
217.196	sterane fragment	191.180	triterpene fragment
372.360	C_{27} steranes	370.359	C_{27} triterpenes
386.391	C_{28} steranes	384.374	C_{28} triterpenes
400.407	C_{29} steranes	398.390	C_{29} triterpenes
414.422	C_{30} steranes	412.406	C_{30} triterpenes
		426.421	C_{31} triterpenes
		440.437	C_{32} triterpenes
		454.453	C_{33} triterpenes
		468.468	C_{34} triterpenes
		482.484	C_{35} triterpenes

[a] For MRM measurements, each of the molecular ions listed are monitored for the transitions to *m/z* 217 for the steranes and *m/z* 191 for the triterpenes.

[b] Atomic mass unit.

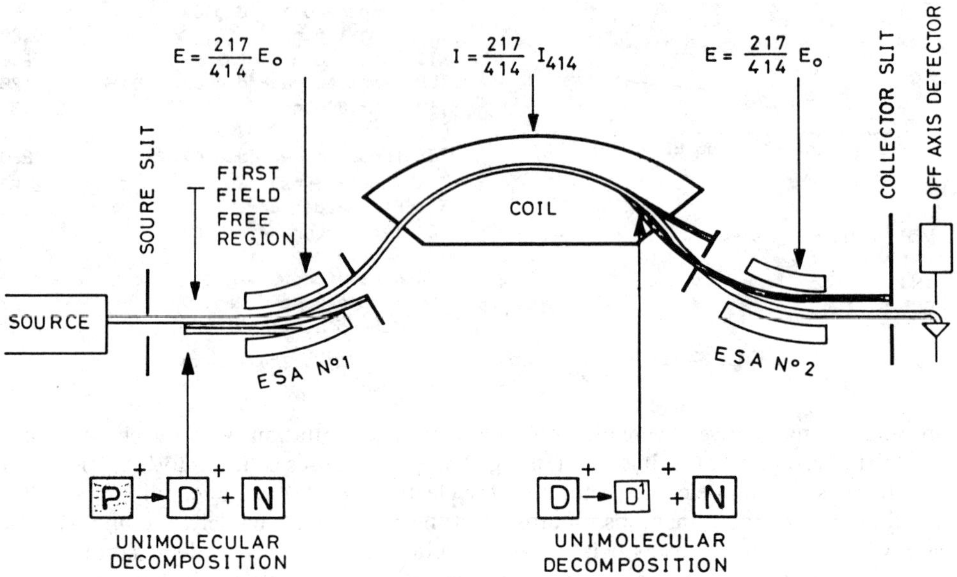

FIG. 2—*Typical analyzer configuration and beam pathways for single reaction,* m/z *414–217.*

Results and Discussion

A mixture of two standards are used to establish the nature of their electron impact (EI) spectra, their unimolecular decomposition spectra, and the expected detection limits for the three methods to be used. The two compounds are, 5α-cholestane and 5β-cholestane (Sigma Chemical Company, St. Louis, Missouri). Full-scan spectra are shown in Fig. 3. Unimolecular daughter spectra are shown in Fig. 4.

The EI spectra show the similarity expected for the two isomers and the base peak at *m/z* 217 that is used to monitor the entire class of compounds, the daughter spectra essentially contain two peaks, one a loss of methyl and the other the *m/z* 217 fragment. The *m/z* 217 fragment is chosen for reaction monitoring as most of the constituents of the oils are *n*-alkanes that readily lose methyl groups and could cause interferences if these ions were monitored. A mixture containing 13 pg of 5α and 7 pg of 5β-cholestane is then injected with the system set to monitor ions in each of the three modes of interest, the resulting chromatograms are shown in Fig. 5. The absolute responses obtained for the detection limit measurement are in the ratio of 100:26:5 for, low-resolution SIR:medium-resolution SIR:MRM. However, the signal to noise ratio is similar for the first two methods and only a factor of two less for the MRM case. The presence of some background in the low-resolution measurement results in the lower sensitivity methods yielding comparable *S/N* ratios. All three methods demonstrate their capability of detecting these substances in the low picogram range.

Full-scan chromatograms of the two oils chosen are shown in Fig. 6. The area in which the biomarkers elute is highlighted. Steranes and triterpenes have been enhanced in the marine oil through crude oil alteration processes. In contrast, the terrestrial oil is an unaltered mature crude typified by abundant "paraffinic" hydrocarbons and by sparse biological marker compounds.

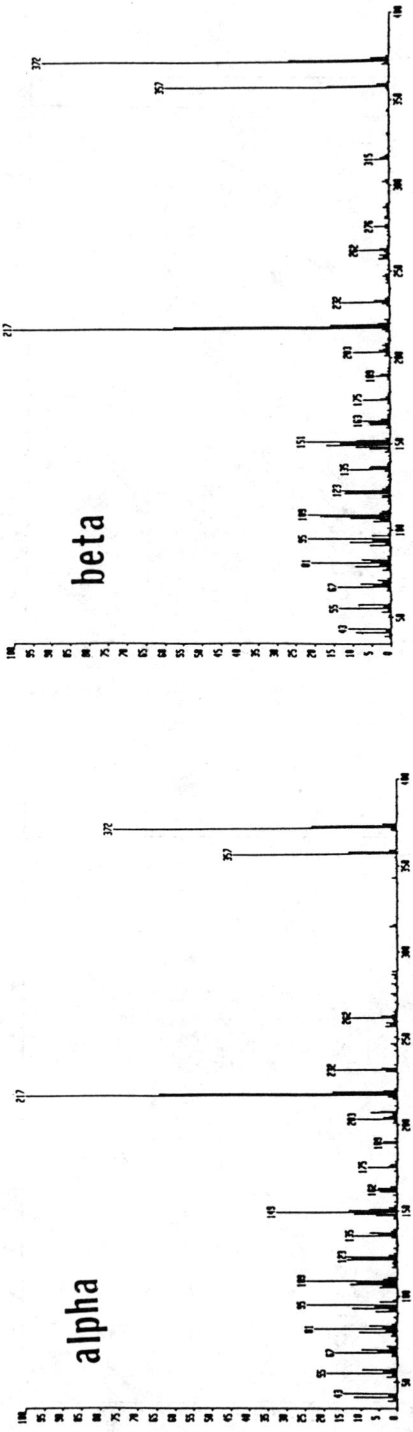

FIG. 3—*Full scan EI spectra of 5α and 5β cholestane.*

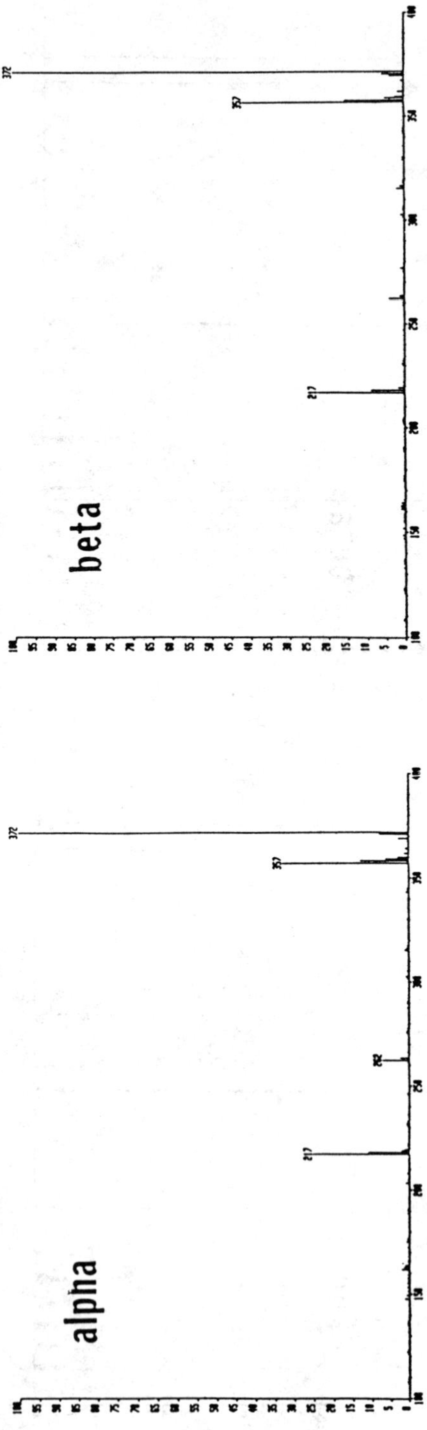

FIG. 4—*Unimolecular daughter spectra of 5α and 5β cholestane.*

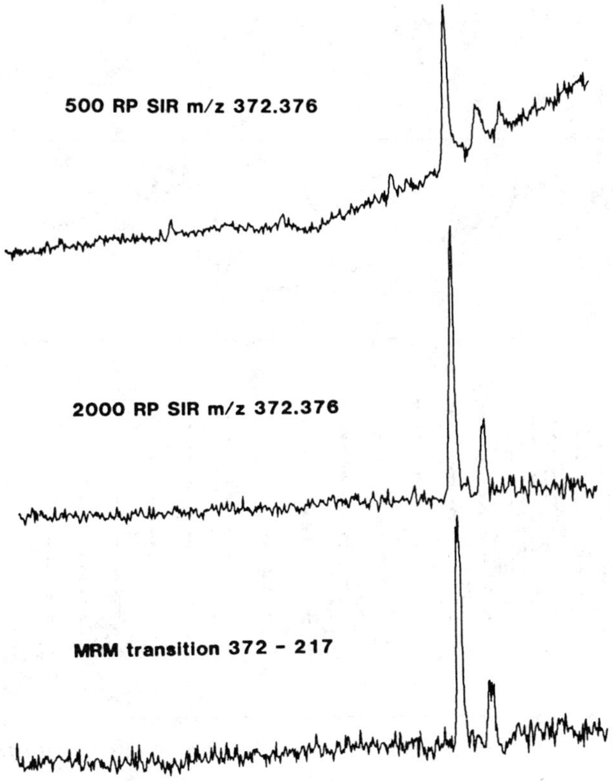

FIG. 5—*Responses obtained for low-level cholestane mixture.*

The results obtained for each of the oils in each of the three modes are shown in the subsequent figures. The sterane and triterpene results are separated to facilitate comparison, although all data are acquired within single runs. Identification of steranes and triterpenes is noted by numbers and letters above the peaks in the MRM traces only. The compound identification is given in Tables 2 and 3.

The steranes in the marine oil are complex in nature and, especially for the C_{30} components, are hard to identify and quantitate in the low and medium-resolution SIR results. The MRM responses are easy to identify and are free from interferences. Due to the large concentration, the *S/N* ratio for each of the modes is equivalent. Figures 7*a* through *d* show the sterane responses in marine oil.

In the case of the terrestrial oil, the sterane concentrations are low (see Figs. 8*a* through *d*). The low and medium-resolution SIR traces are characterized by high background, making identification and quantification difficult. The MRM traces show good *S/N* and low interference. As may be expected for such a terrestrial oil, there is insignificant input from paleo marine organic matter, signified by negligible response for C_{30} components [*11*].

The triterpenes are present in large concentrations within the marine sample (see Figs. 9*a* through *d*). However, the selectivity advantage of the MRM method is again evident. The baseline rise seen in the SIR traces is due to unresolved aromatic components, this is not seen to the same degree in MRM mode.

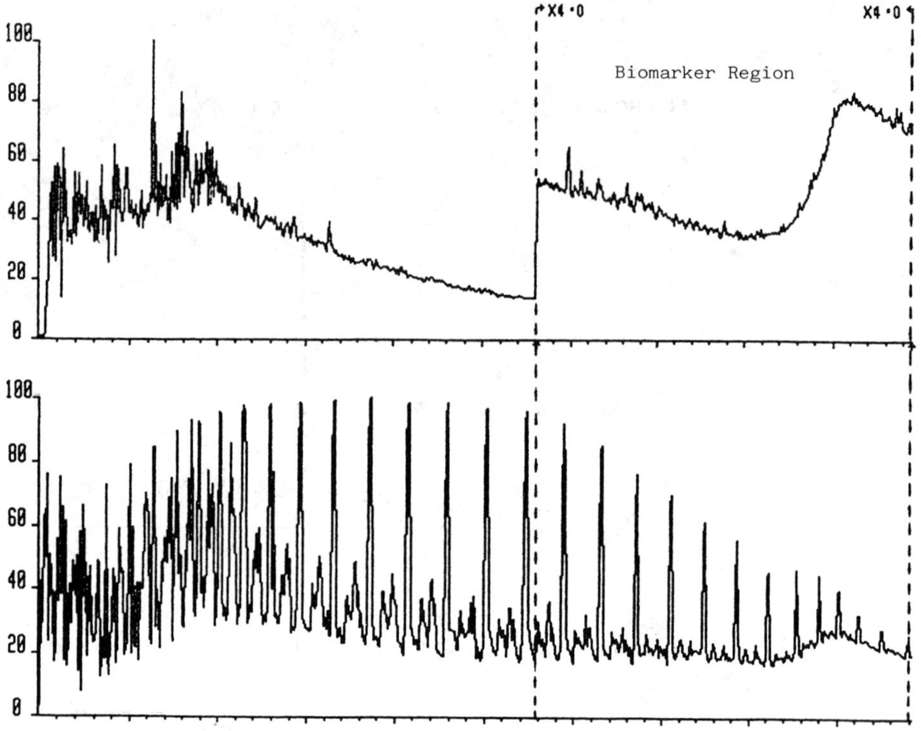

FIG. 6—*Total ion current traces for the marine source oil* (top) *and terrestrial source* (bottom).

TABLE 2—*Sterane peak identifications.*

Peak No.	Identification	Peak No.	Identification
1	13β,17α-diacholestane (20S)	17	13β,17α-24-ethylcholestane (20S)
2	13β,17α-diacholestane (20R)	18	13α,17α-24-ethylcholestane (20R)
3	13α,17β-diacholestane (20S)	19	13α,17β-24-ethylcholestane (20S)
4	13α,17β-diacholestane (20R)	20	13α,17β-24-ethylcholestane
5	5α,14α-17α-cholestane (20S)	21	5α,14α,17α-24-ethylcholestane
6	5α,14β,17β-cholestane (20R)	22	5α,14β,17β-24-ethylcholestane
7	5α,14β,17β-cholestane (20S)	23	5α,14β,17β-24-ethylcholestane
8	5α,14α,17α-cholestane (20R)	24	5α,14α,17α-24-ethylcholestane
9	13β,17α,-24-methyldiacholestane (20S)	25	13β,17α,-4-desmethyl-C_{30} sterane
10	13β,17β,-24-methyldiacholestane (20R)	26	13β,17α,-4-desmethyl-C_{30} sterane
11	13α,17β,-24-methyldiacholestane (20S)	27	13α,17β,-4-desmethyl-C_{30} sterane
12	13α,17β,-24-methyldiacholestane (20R)	28	13α,17β,-4-desmethyl-C_{30} sterane
13	5α,14α,17α-24-methylcholestane (20S)	29	5α,14α,17α-4-desmethyl-C_{30} sterane
14	5α,14β,17β-24-methylcholestane (20R)	30	5α,14β,17β-4-desmethyl-C_{30} sterane
15	5α,14β,17β-24-methylcholestane (20S)	31	5α,14β,17β-4-desmethyl-C_{30} sterane
16	5α,14α,17α-24-methylcholestane (20R)	32	5α,14α,17α-4-desmethyl-C_{30} sterane

TABLE 3—*Triterpene identifications.*

Peak Letter	Identification	Peak Letter	Identification
A	18α,21β trisnorhopane (T_s)	G	17α,21β homohopane (22S)
B	17α,21β trisnorhopane (T_m)	H	17α,21β homohopane (22R)
Z	17α,18α bisnorhopane	J	17α,21β bisnorhopane (22S + R)
C	17α,21β norhopane	K	17α,21β trisnorhopane (22S + R)
C_1	C_{29} pentacyclic triterpenoid	L	17α,21β tetrakishomohopane (22S + R)
E	17α,21β hopane	M	17α,21β pentakishomohopane (22S + R)

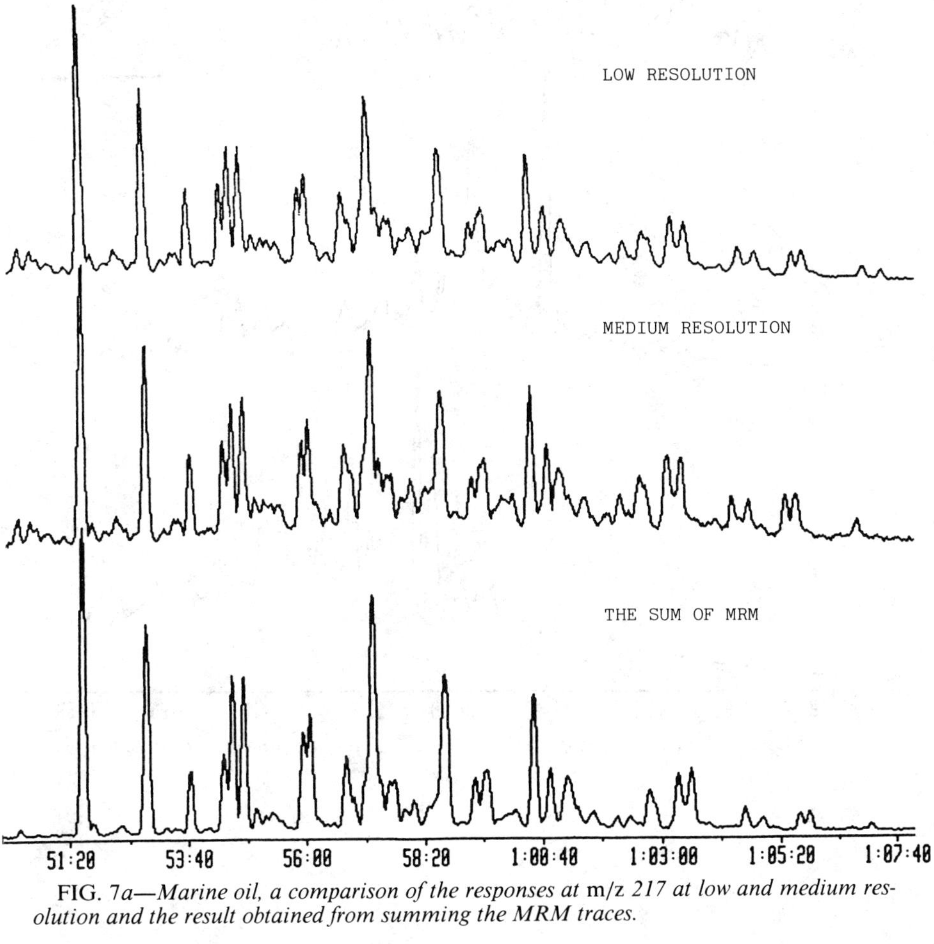

FIG. 7*a*—*Marine oil, a comparison of the responses at* m/z *217 at low and medium resolution and the result obtained from summing the MRM traces.*

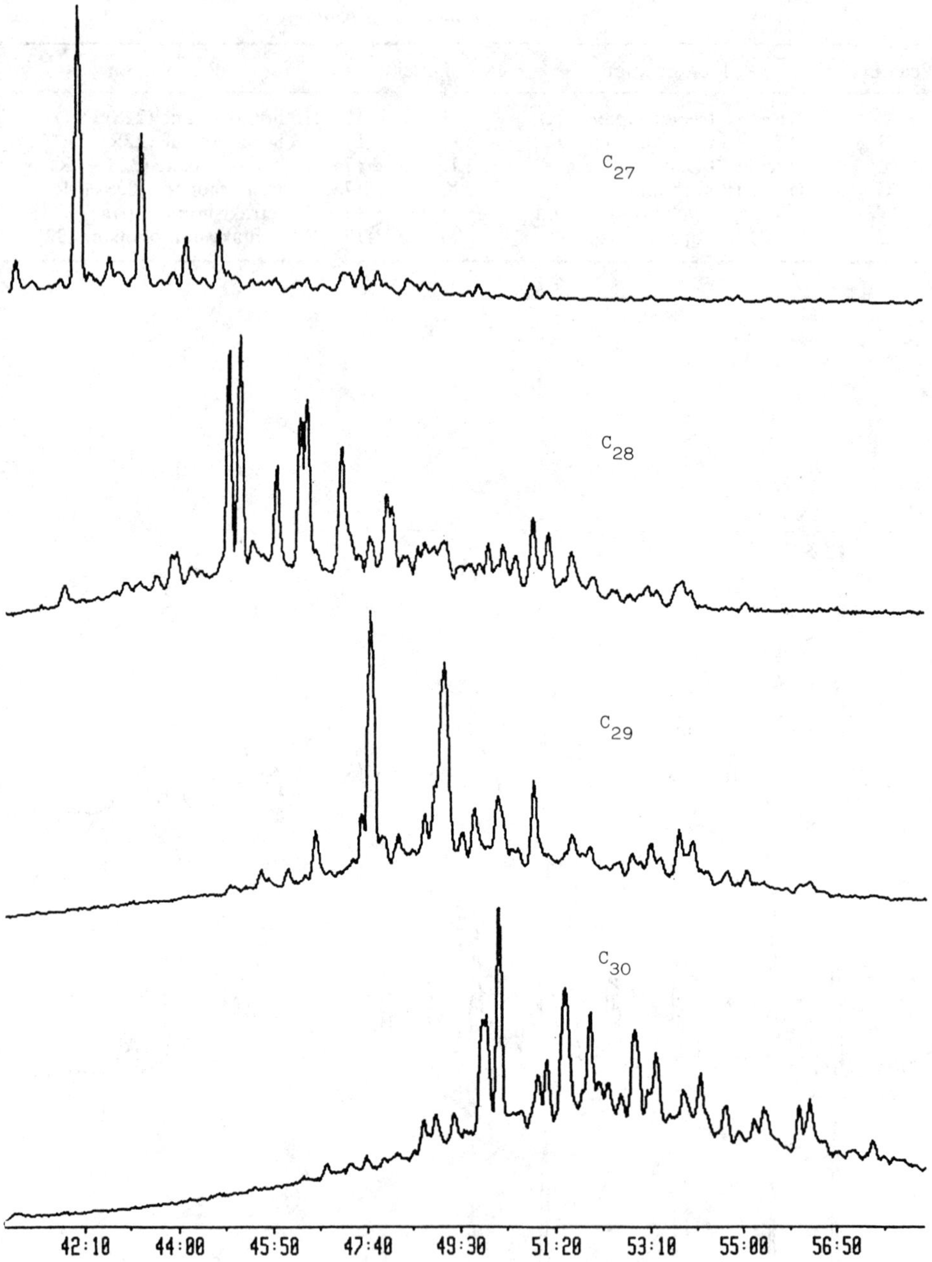

FIG. 7b—*Marine oil, responses for C_{27} to C_{30} steranes, low-resolution SIR.*

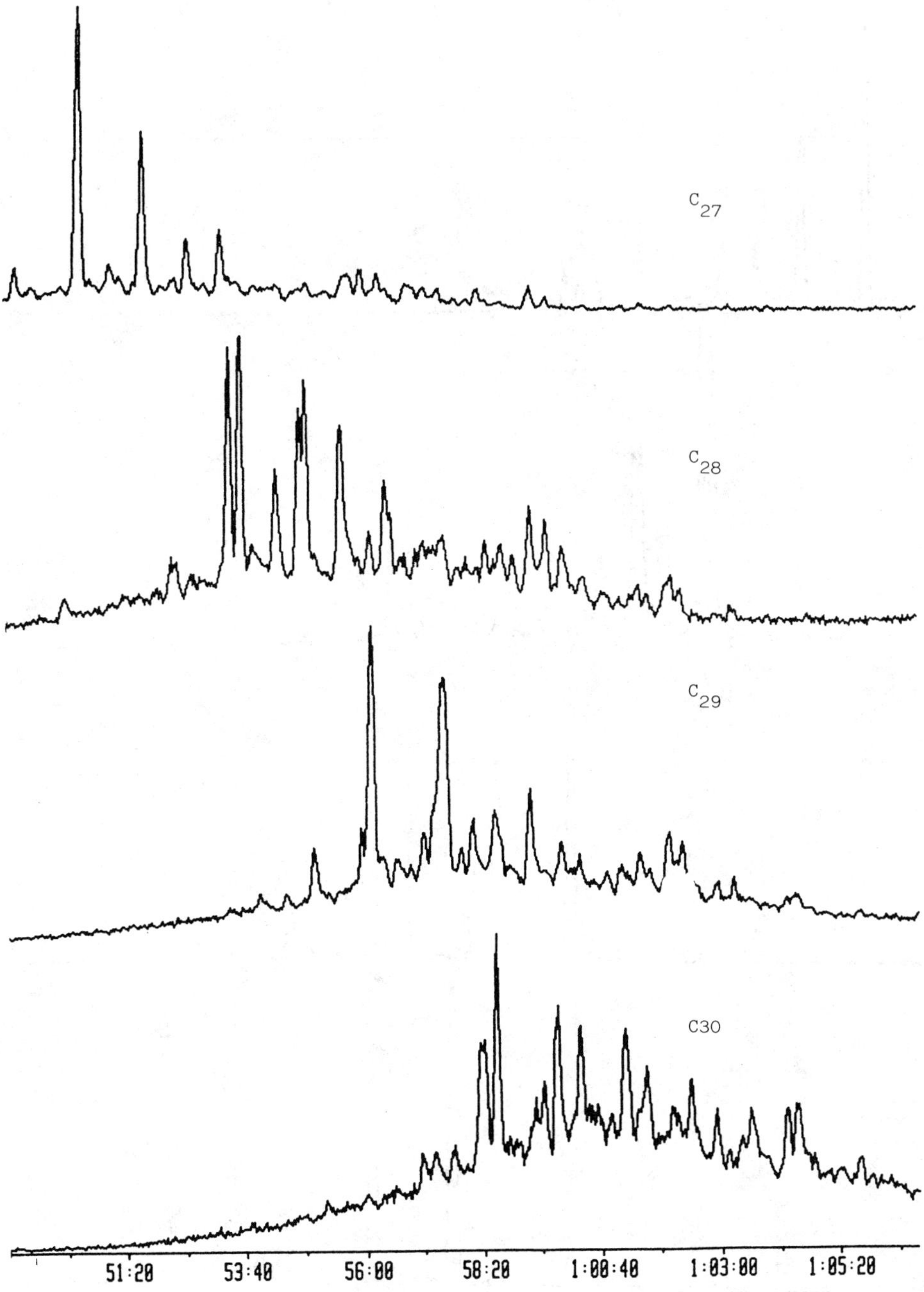

FIG. 7c—*Marine oil, responses for C_{27} to C_{30} steranes, medium-resolution SIR.*

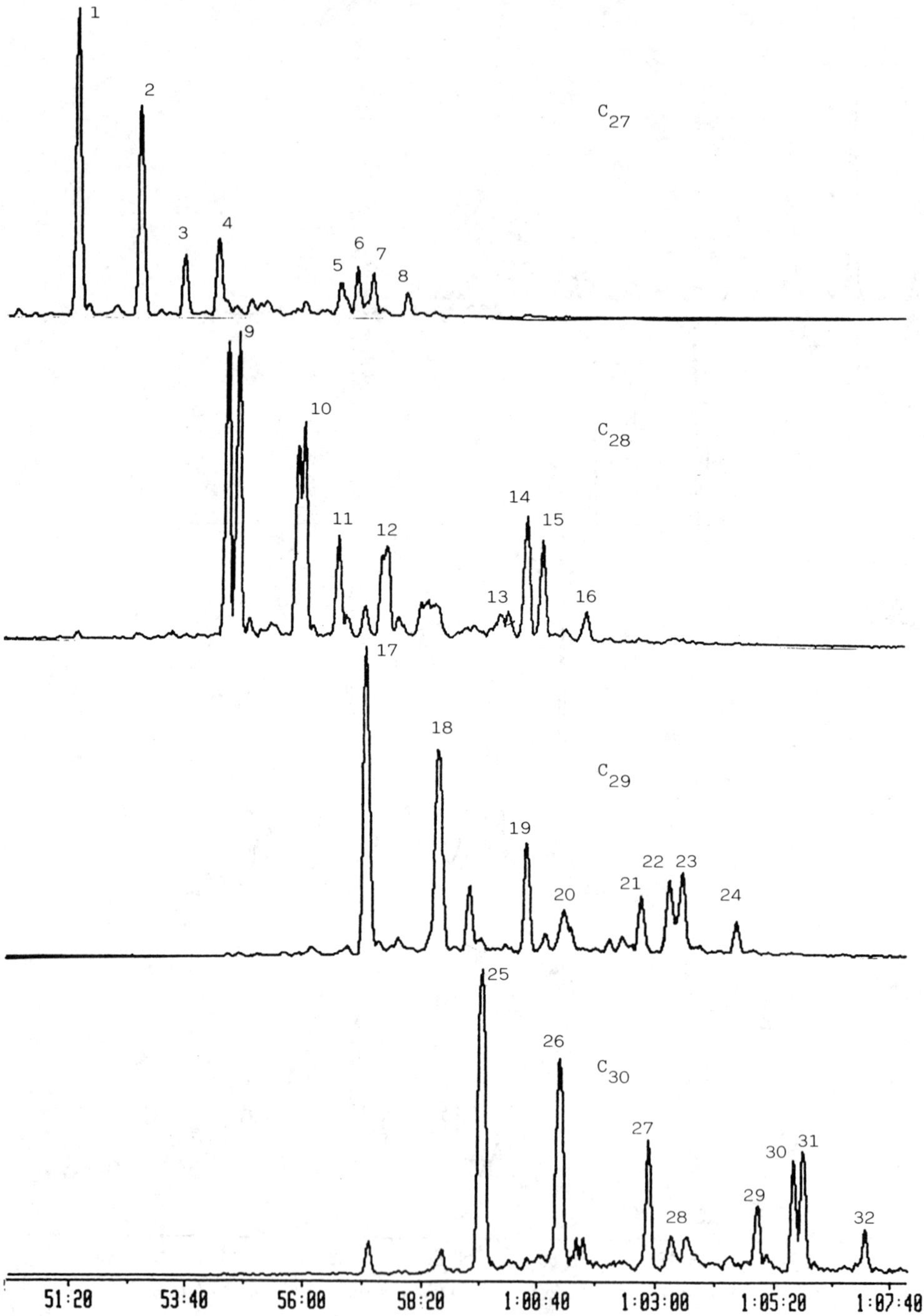

FIG. 7*d*—*Marine oil, responses for C_{27} to C_{30} steranes, MRM.*

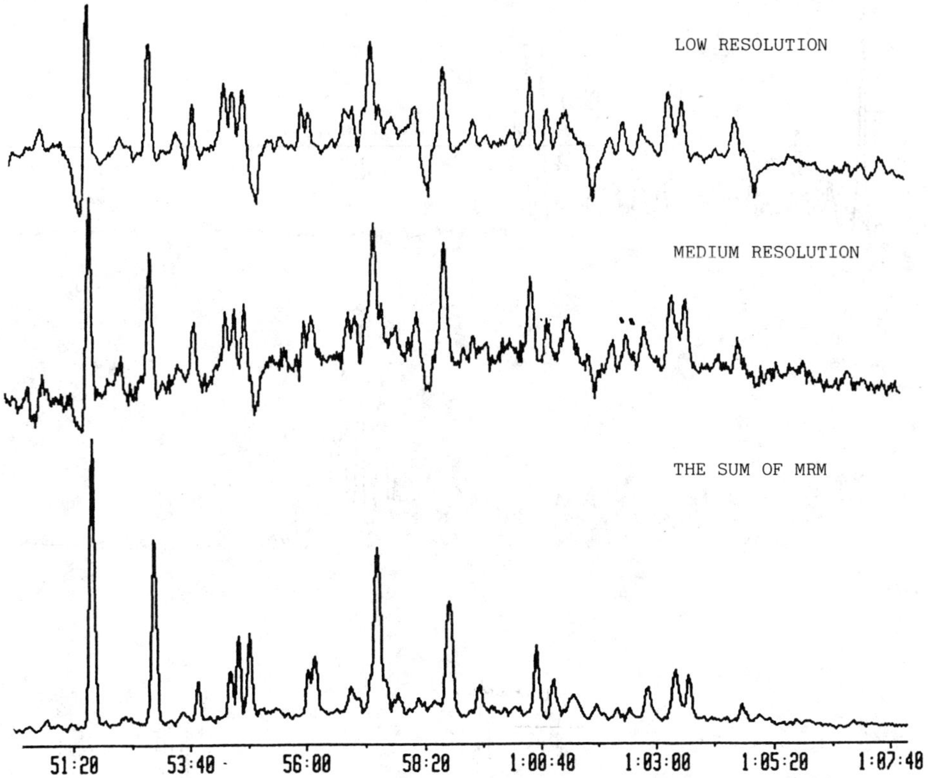

FIG. 8*a*—*Terrestrial oil, a comparison of the responses at* m/z *217 at low and medium resolution and the result obtained from summing the MRM traces.*

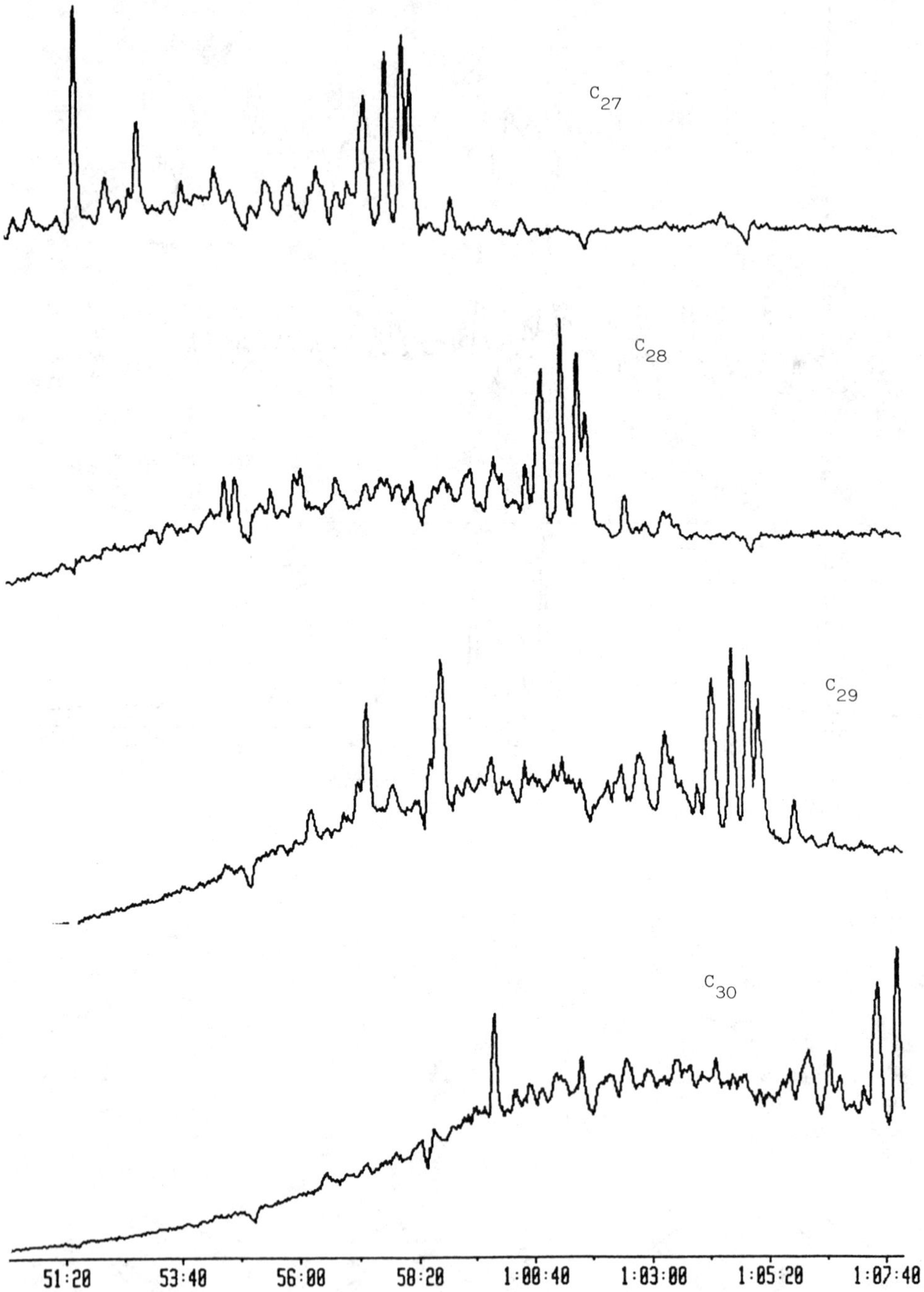

FIG. 8*b*—*Terrestrial oil, responses obtained for* C_{27} *to* C_{30} *steranes, low-resolution SIR.*

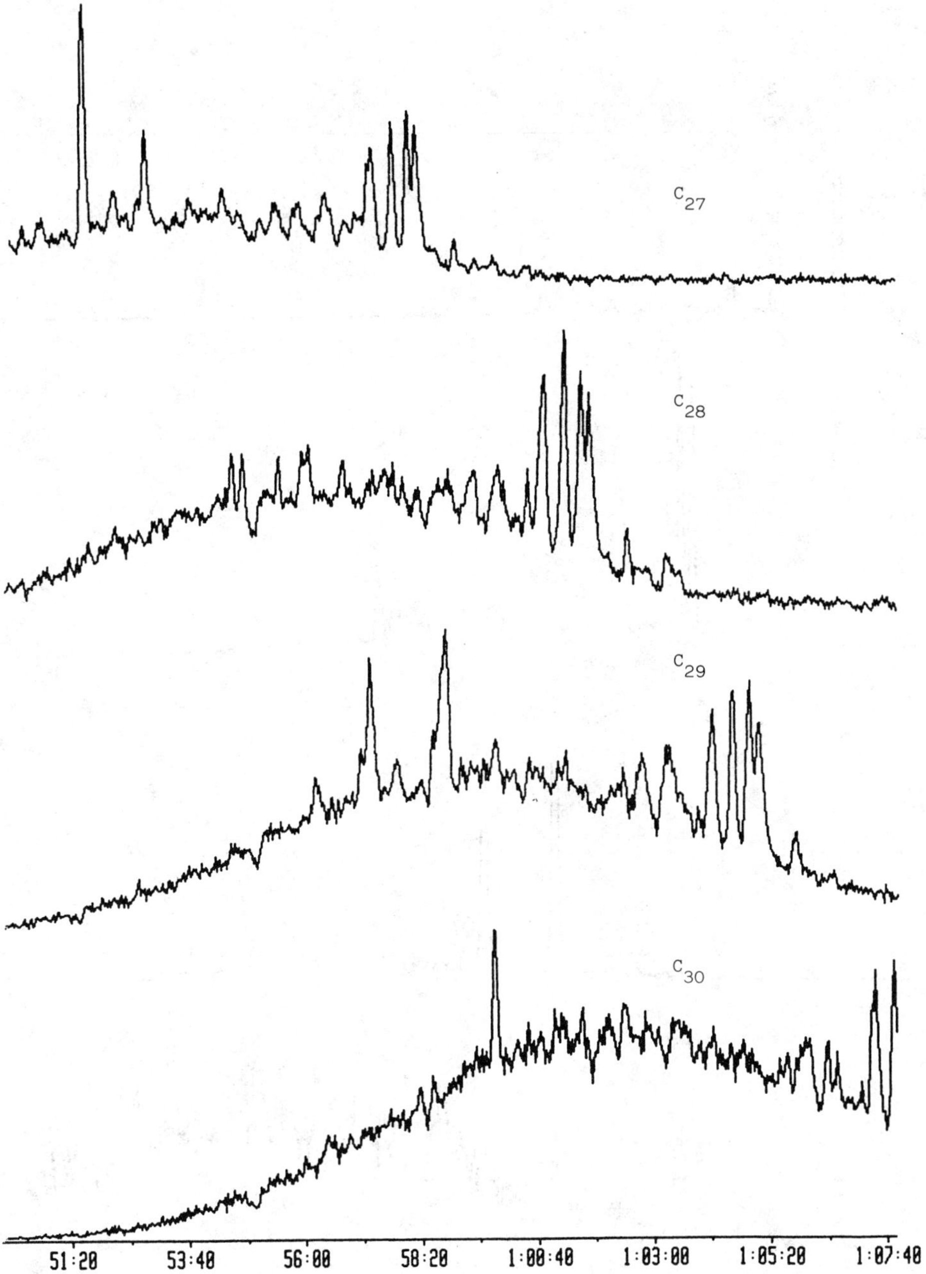

FIG. 8c—*Terrestrial oil, responses obtained for C_{27} to C_{30} steranes, medium-resolution SIR.*

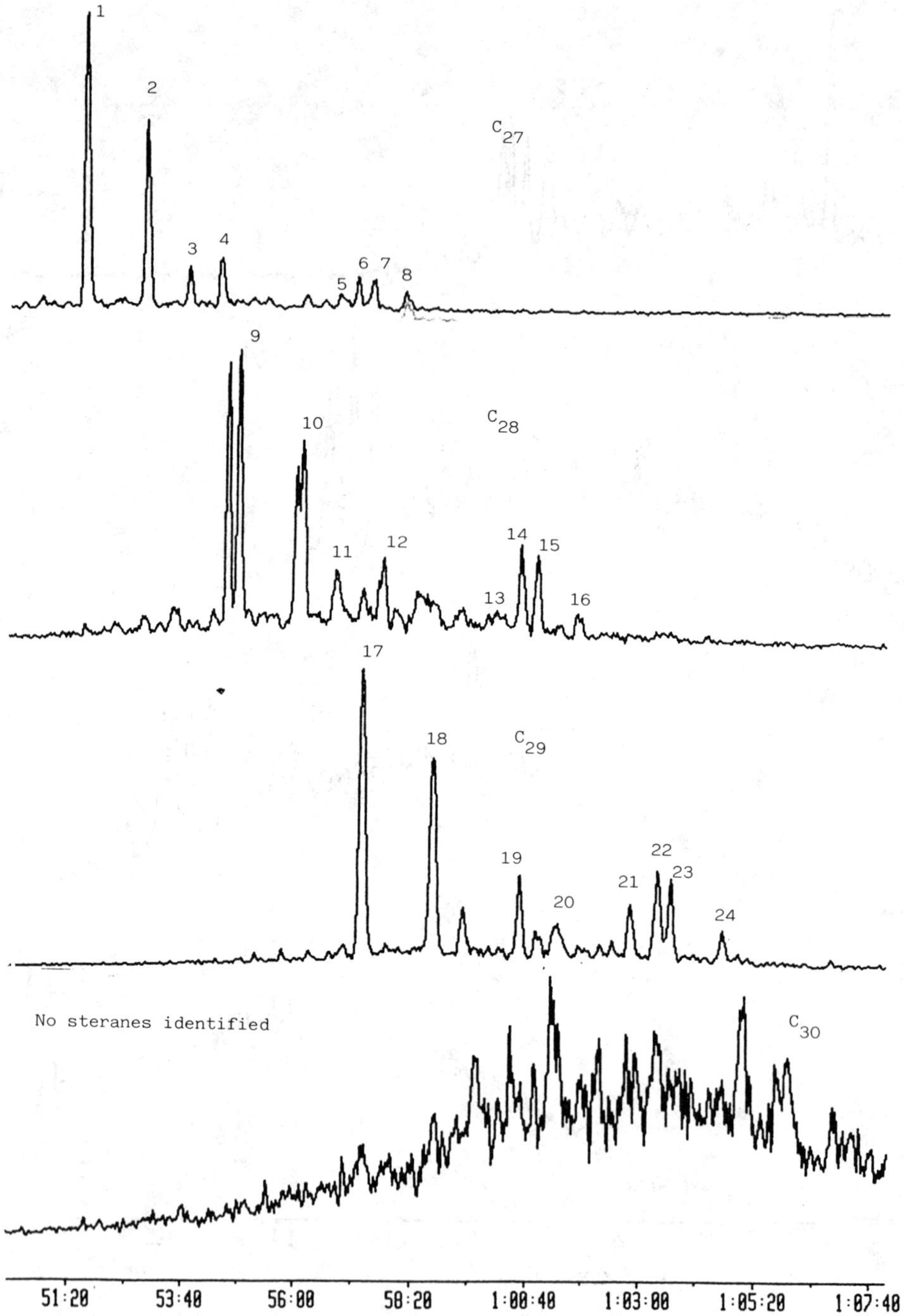

FIG. 8*d*—*Terrestrial oil, responses obtained for* C_{27} *to* C_{30} *steranes, MRM.*

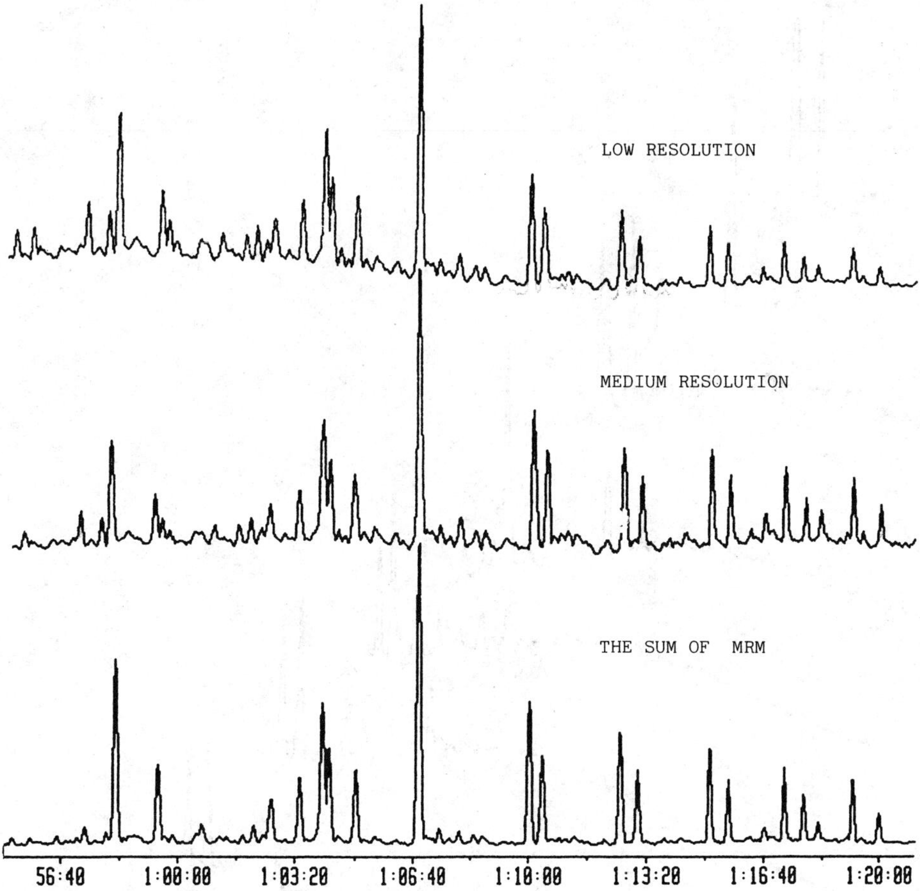

FIG. 9*a*—*Marine oil, a comparison of the responses at* m/z *191 at low and medium resolution and the result obtained from summing the MRM traces.*

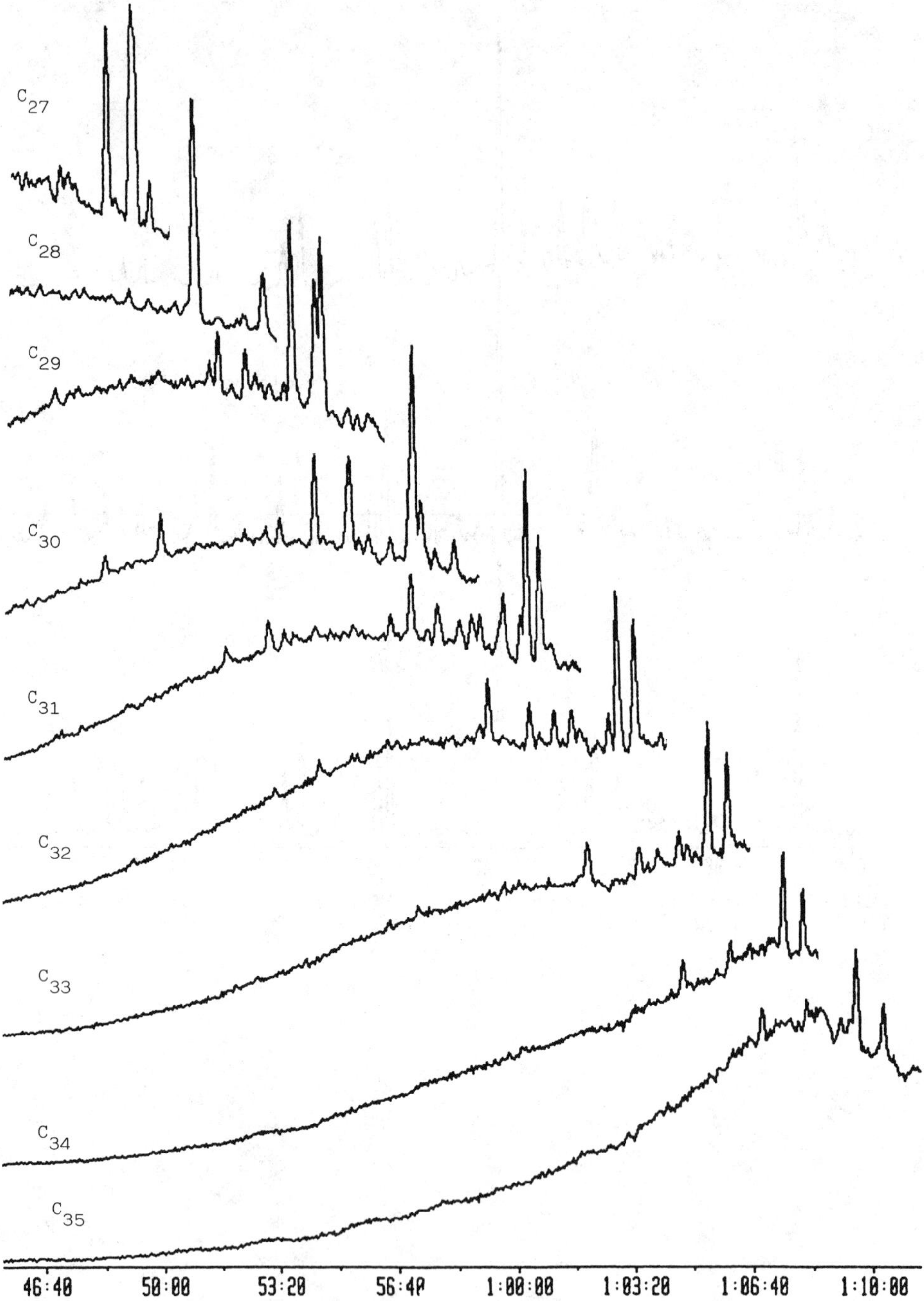

FIG. 9*b*—*Marine oil, responses obtained for* C_{27} *to* C_{35} *triterpenes, low-resolution SIR.*

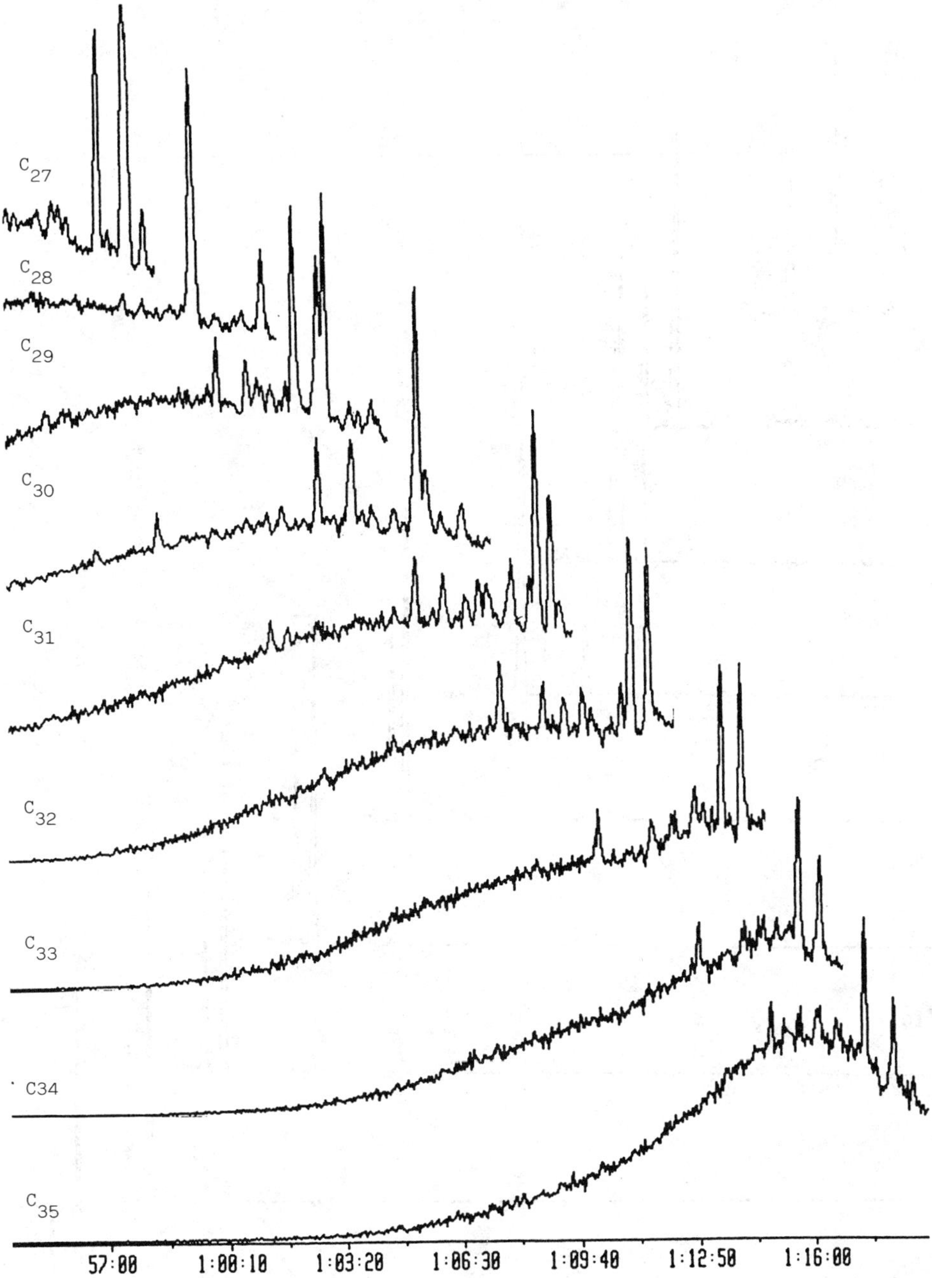

FIG. 9c—*Marine oil, responses obtained for* C_{27} *to* C_{35} *triterpenes, medium-resolution SIR.*

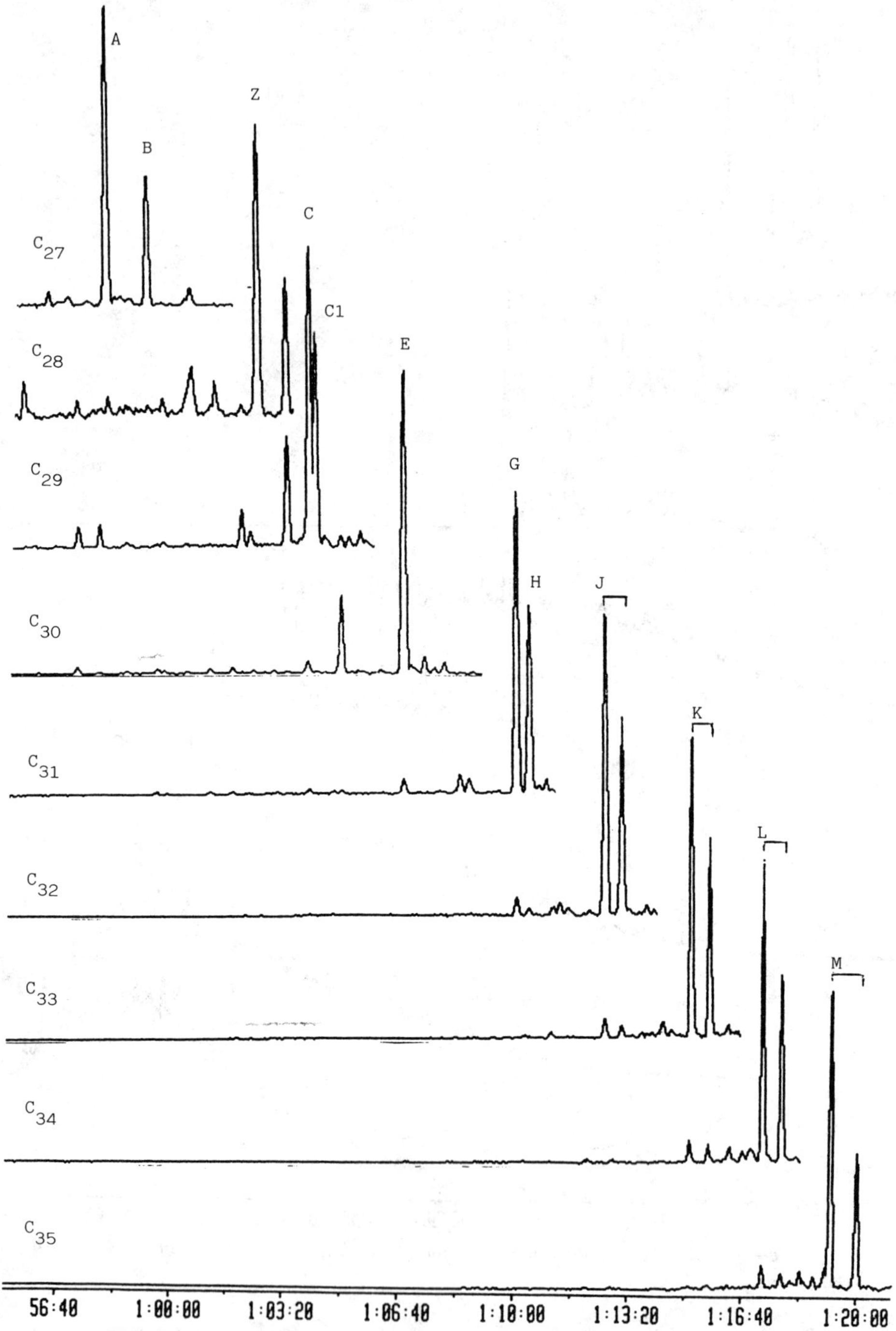

FIG. 9*d*—*Marine oil, responses obtained for C_{27} to C_{35} triterpenes, MRM.*

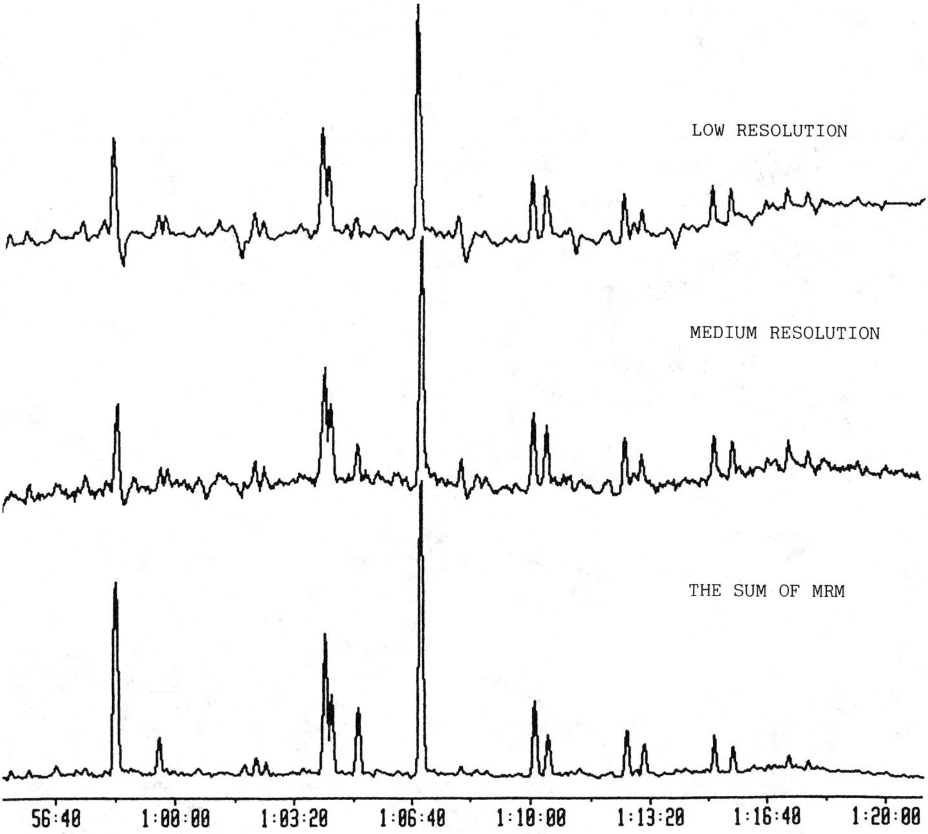

FIG. 10*a*—*Terrestrial oil, a comparison of the responses at* m/z *191 at low and medium resolution and the result obtained from summing the MRM traces.*

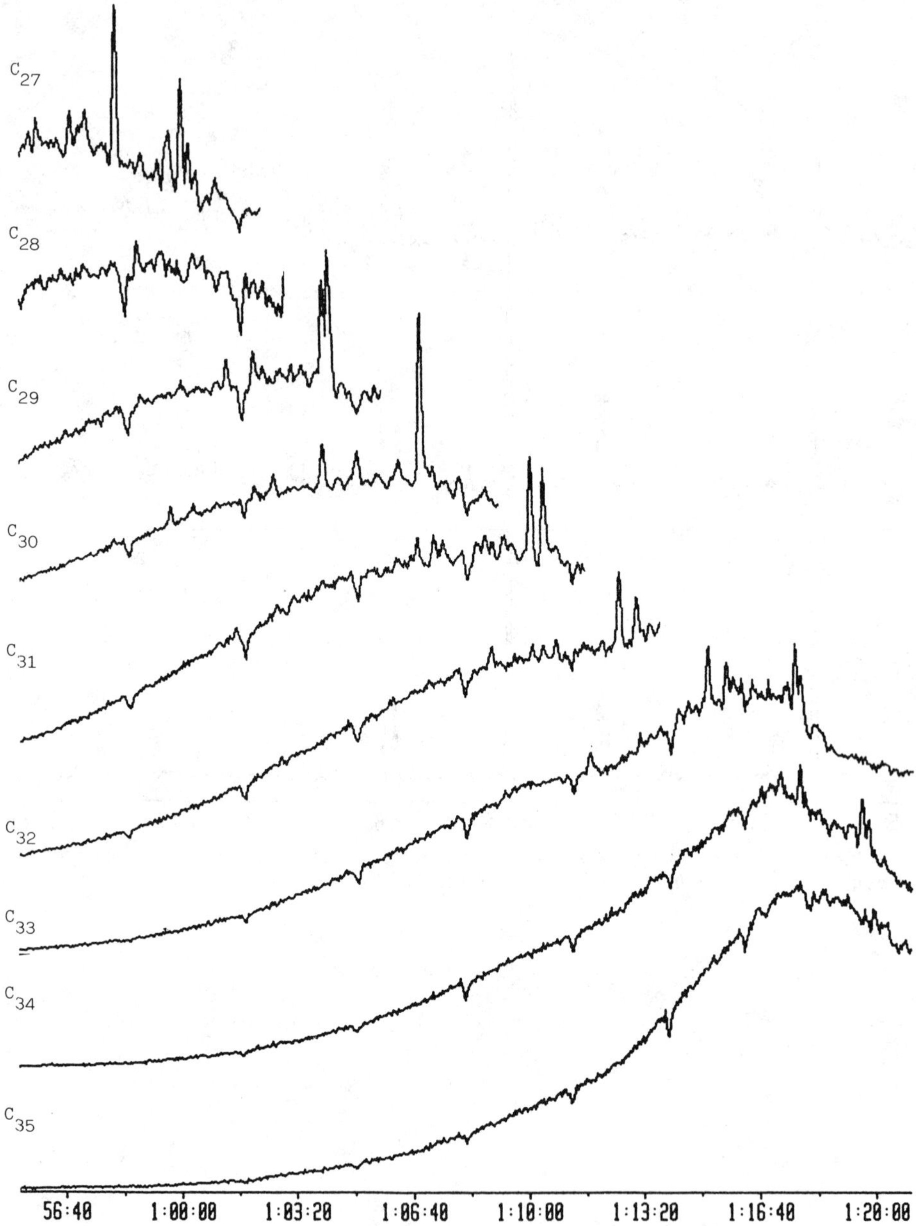

FIG. 10*b*—*Terrestrial oil, responses obtained for C_{27} to C_{35} triterpenes, low-resolution SIR.*

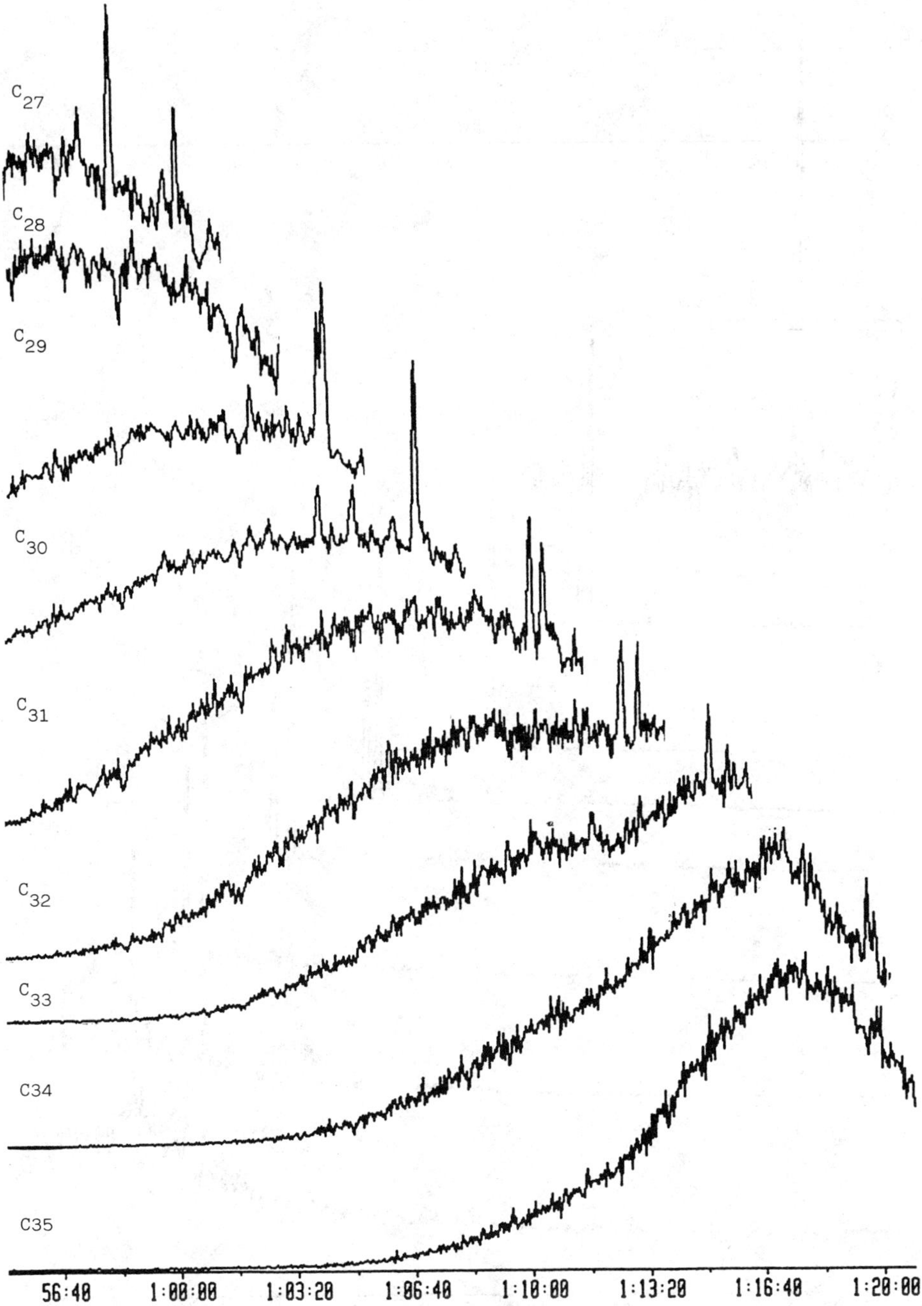

FIG. 10*c*—*Terrestrial oil, responses obtained for C_{27} to C_{35} triterpenes, medium-resolution SIR.*

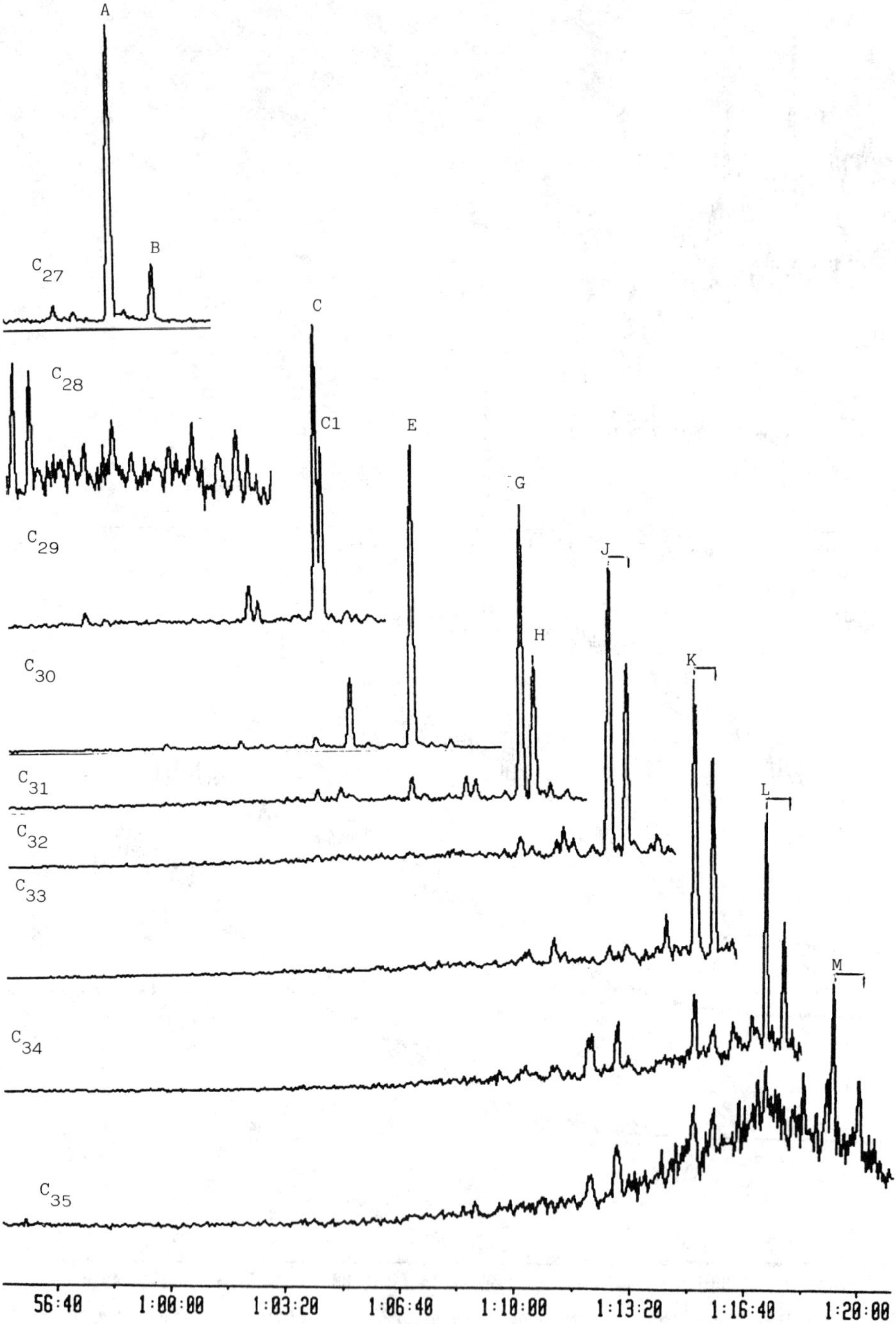

FIG. 10*d*—*Terrestrial oil, responses obtained for C_{27} to C_{35} triterpenes, MRM.*

In the case of triterpenes in the terrestrial oil, higher homologue hopanes are sparse and are difficult to identify (see Figs. 10*a* through *d*). The presence of the background prohibits the detection of components beyond C_{32} at low resolution and C_{33} at medium resolution. In MRM mode components are seen to C_{35}. Despite the absolute signal level being less for the MRM, it clearly yields a lower detection limit in a real sample.

Conclusion

The data sets illustrate the advantage of MRM for the analysis of biomarkers in *whole* oils. The perceived disadvantage in sensitivity from the measurements using standards is more than compensated for by the better selectivity obtained. This selectivity is important even when sensitivity is not an issue, as with the marine oil when evaluating the C_{30} components. The SIR traces suffer from too much interference to allow confident identification and quantification. The increase in resolving power in SIR has little advantage for this analysis, although related measurements of, say, monoaromatic steranes are significantly improved where the interferences are alkanes themselves [*8*].

References

[*1*] Seifert, W. K. and Moldowan, J. M., "Paleoreconstruction by Biological Markers," *Geochimica et Cosmochimica Acta,* Vol. 45, 1981, pp. 783–794.

[*2*] Seifert, W. K., and Moldowan, J. M., "Applications of Steranes and Monoaromatics to the Maturation, Migration and Source of Crude Oils," *Geochimica et Cosmochimica Acta,* Vol. 42, 1978, pp. 77–95.

[*3*] Mackenzie, A. S., Brassell, S. C., Eglinton, G., and Maxwell, J. R., "Chemical Fossils: The Geological Fate of Steroids," *Science,* Vol. 217, 1982, pp. 491–504.

[*4*] Peters, K. E., Moldowan, J. M., Schoell, M., and Hempkins, W. B., "Petroleum Isotopic and Biomarker Composition Related to Source Rock Organic Matter and Depositional Environment," *Organic Geochemistry,* Vol. 10, 1986, p. 1727.

[*5*] Bjoroy, M., Brooks, P. W., and Hall, K., "An Oil/Oil Correlation Study Utilising High Resolution GC-MS," *Advances in Organic Geochemistry,* 1981, pp. 87–93.

[*6*] Meyer, T. and Christie, O. H. J., "Aspects of Biomarker Analysis by GC-MS with Selective Metastable Ion Monitoring," *Analytica Chimica Acta,* Vol. 161, 1984, pp. 65–74.

Steen, A., "GC/MS Analysis of C_{27}–C_{30} Steranes," *Organic Geochemistry,* Vol. 10, 1986, pp. 1137–1142.

[*8*] Mackenzie, A. S., Disko, U., and Rullkotter, J., "1983 Determination of Hydrocarbon Distributions in Oils and Sediment Extracts by GC-High Resolution Mass Spectrometry," *Organic Geochemistry,* Vol. 5, 1983, pp. 57–63.

[*9*] Warburton, G. A. and Zumberge, J. E., "Determination of Petroleum Sterane Distributions by Mass Spectrometry with Selective Metastable Ion Monitoring," *Analytical Chemistry,* Vol. 55, 1983, pp. 123–126.

[*10*] Bateman, R. H., Burns, P., Owen, R., and Parr, V. C., "A Novel General Purpose Double Focussing Mass Spectrometer," presented at the 23rd Meeting, American Society for Massed Spectrometry, San Diego, 1985.

[*11*] Moldowan J. M. et al., "C_{30}-steranes, Novel Markers for Marine Petroleums and Sedimentary Rocks," *Cosmochimica Acta,* Vol. 48, 1984, pp. 2767–2768.

Brian D. Beato,[1] Richard A. Yost,[1] and J. Martin E. Quirke[2]

The Effect of Metal Atoms on Electron Ionization and Ammonia Chemical Ionization Mass Spectra of Metalloporphyrins: Implications for Geoporphyrin Analysis

REFERENCE: Beato, B. D., Yost, R. A., and Quirke, J. M. E., **"The Effect of Metal Atoms on Electron Ionization and Ammonia Chemical Ionization Mass Spectra of Metalloporphyrins: Implications for Geoporphyrin Analysis,"** *Novel Techniques in Fossil Fuel Mass Spectrometry, ASTM STP 1019,* T. R. Ashe and K. V. Wood, Eds., American Society for Testing and Materials, Philadelphia, 1989, pp. 84–93.

ABSTRACT: Eight metal complexes of octaethylporphyrin were analyzed by electron ionization (EI) and ammonia chemical ionization (NH_3 CI) mass spectrometry. In both modes, features of the mass spectra of the metal complexes correlated well with the stability index of the metalloporphyrin, $S_i = [100(PE)(V)]/R$, where PE is the Pauling Electronegativity of the metal, V is the valence of the metal, and R is the ionic radius of the metal. Metalloporphyrins with high stability indices produced abundant doubly charged fragments in the EI mode, and gave very little fragmentation with NH_3 CI. Metalloporphyrins with low stability indices gave less intense doubly charged ions in EI, but gave abundant monopyrrolic fragments, and some dipyrrolic fragments under NH_3 CI conditions. These results will be very valuable in geoporphyrin analysis, since it will now be possible to control the degree of fragmentation by inserting an appropriate metal into demetalated geoporphyrins.

KEY WORDS: metal complexes, octaethylporphyrin, electron ionization, ammonia chemical ionization, mass spectrometry, stability index, doubly charged ions, monopyrrolic fragments, dipyrrolic fragments, demetalated geoporphyrins, fossil fuels

Geoporphyrins are biological marker compounds that can provide valuable information on fossil fuels, such as sample maturation, depositional environment of sample, and oil-oil, and oil-source rock correlation [*1,2*]. Structure elucidation of metalloporphyrins by mass spectrometry is of considerable interest in fossil fuel research as this is the first stage in determining the full value of these compounds for oil exploration. Electron ionization (EI) mass spectral analysis offers structural information including the molecular weight of these compounds, and identification of some of the peripheral substituents [*3–5*]. It does not, however, indicate the structures or the sequence of the individual pyrrolic rings that make up the geoporphyrin. Methane chemical ionization (CI) mass spectral analysis typically provides strictly molecular weight information about the intact molecular ion. The use of ammonia CI (NH_3 CI) is valuable in structure elucidation of metalloporphyrins because it fragments the metalloporphyrin into pyrrolic ions, thus helping to reveal the

[1] Research assistant and associate professor, respectively, Chemistry Department, University of Florida, Gainesville, FL 32611.

[2] Associate professor, Department of Chemistry, Florida International University, Miami, FL 33199.

structures of the pyrroles that make up the metalloporphyrin, as well as indicating which substituents reside on which pyrroles [6,7]. Thus, NH_3 CI is a good compliment to EI in the determination of geoporphyrin structures by mass spectrometry.

Structure elucidation of metalloporphyrins is enhanced by the use of tandem mass spectrometry (MS/MS), in which mass-selected ions are passed through a collision cell containing an inert gas and undergo collisionally activated dissociation (CAD) [8]. Mass analysis of the fragments of those ions initially formed by EI or CI helps identify the original compound. One drawback to the use of MS/MS to obtain more structural information is the requirement for larger amounts of compound, especially when the ions of interest are of low relative abundance in the original mass spectrum. This can pose a problem in geoporphyrin analyses, since some of the most useful MS/MS data would be those produced from the relatively low abundance of pyrrolic fragments that are generated by NH_3 CI. In single stage mass spectrometry, the only opportunity for varying the amount of fragmentation is in the ion source. Quite often, the compound either fragments too much for adequate molecular weight determination, or too little for structure elucidation. Clearly, it would be very useful to be able to vary the extent of fragmentation to suit specific analytical needs.

In this study, the extent of EI and NH_3 CI fragmentation is shown to depend on the metal atom in the metalloporphyrin. Some metals enhance fragmentation, while others diminish fragmentation, compared to demetalated porphyrins. A systematic characterization of the effect of the metal atoms on EI and NH_3 CI mass spectra of metalloporphyrins should permit the optimum selection of chelated metal to enhance or diminish fragmentation, as desired. Demetalation and metal insertion procedures are usually facile [9].

To determine the effect of various metals on fragmentation, EI and NH_3 CI mass spectral data were obtained for a series of metalloporphyrins. The extent of fragmentation for each metalloporphyrin under both ionization methods was determined and these experimental data were compared to theoretical physical constants in an effort to interpret the observations. The constant to which the data best correlated was a measure of the stability of the metalloporphyrin towards demetalation by acid in solution, known as the stability index [9]. Buchler defined the stability index, S_i, as

$$S_i = \frac{100\,(\mathrm{PE})\,(V)}{R}$$

where

PE = the Pauling Electronegativity of the metal,
V = the valence of the metal, and
R = the ionic radius of the metal.

The lower the stability index, the greater the reactivity of the metalloporphyrin in demetalation reactions. Although originally developed to describe solution chemistry, this index is shown to correlate with features of EI and NH_3 CI mass spectral data as well.

Procedure

Equipment

All spectra were obtained on a Finnigan MAT TSQ45 triple quadrupole mass spectrometer/data system operated in the positive ion mode with samples introduced via a solids probe in which 2 μg of sample were heated at 99°C/min. Normal mass spectra were given by scanning the first quadrupole only and passing all of the ions through the second and third quadrupoles (operated in the radio frequency (rf) only mode).

Electron Ionization Mass Spectra

The EI spectra were obtained with an electron energy of 70 eV, an emission current of 0.3 mA, and an analyzer pressure of 3×10^{-7} torr.

Ammonia Chemical Ionization Mass Spectra

The NH_3 CI spectra were obtained with an electron energy of 100 eV, an emission current of 0.3 mA, a source pressure of 0.4 torr, and an analyzer pressure of 1×10^{-5} torr.

Standards

All of the samples were 99.9% pure, purchased from Aldrich, Milwaukee, WI. The metalloporphyrins studied (Fig. 1) were eight derivatives of 2,3,7,8,12,13,17,18-octaethyl-21H,23H-porphine (OEP): Fe(III)(OEP)Cl, Cr(III)(OEP)Cl, Mn(III)(OEP)Cl,Ni(II)(OEP), Cu(II)(OEP), Co(II)(OEP), Zn(II)(OEP), Mg(II)(OEP).

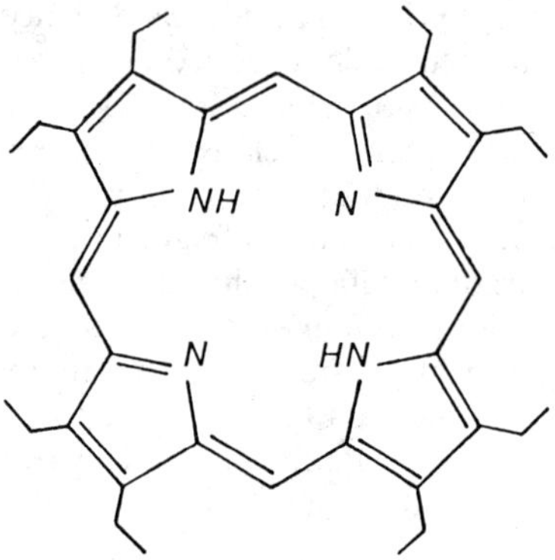

FIG. 1—*Structure of metal-free OEP.*

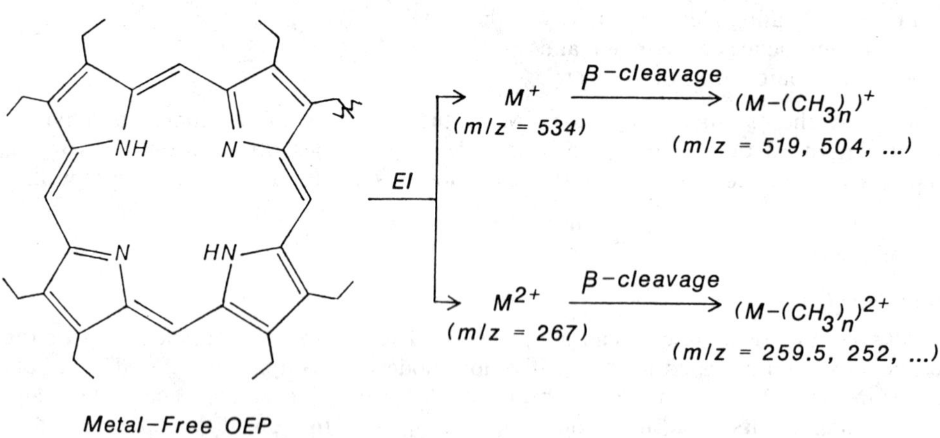

FIG. 2—*Fragmentation pathways of metal-free OEP under EI conditions.*

Results and Discussion

Electron Ionization Mass Spectrometry of Porphyrins

The EI mass spectra of porphyrins are characterized by β-cleavage of peripheral substituents as well as by singly and doubly charged molecular and fragment ions [*3–5*]. Under EI conditions, OEP undergoes β-cleavage of the ethyl substituents, resulting in the loss of methyl fragments, in both the singly and doubly charged regions (Fig. 2). The EI mass

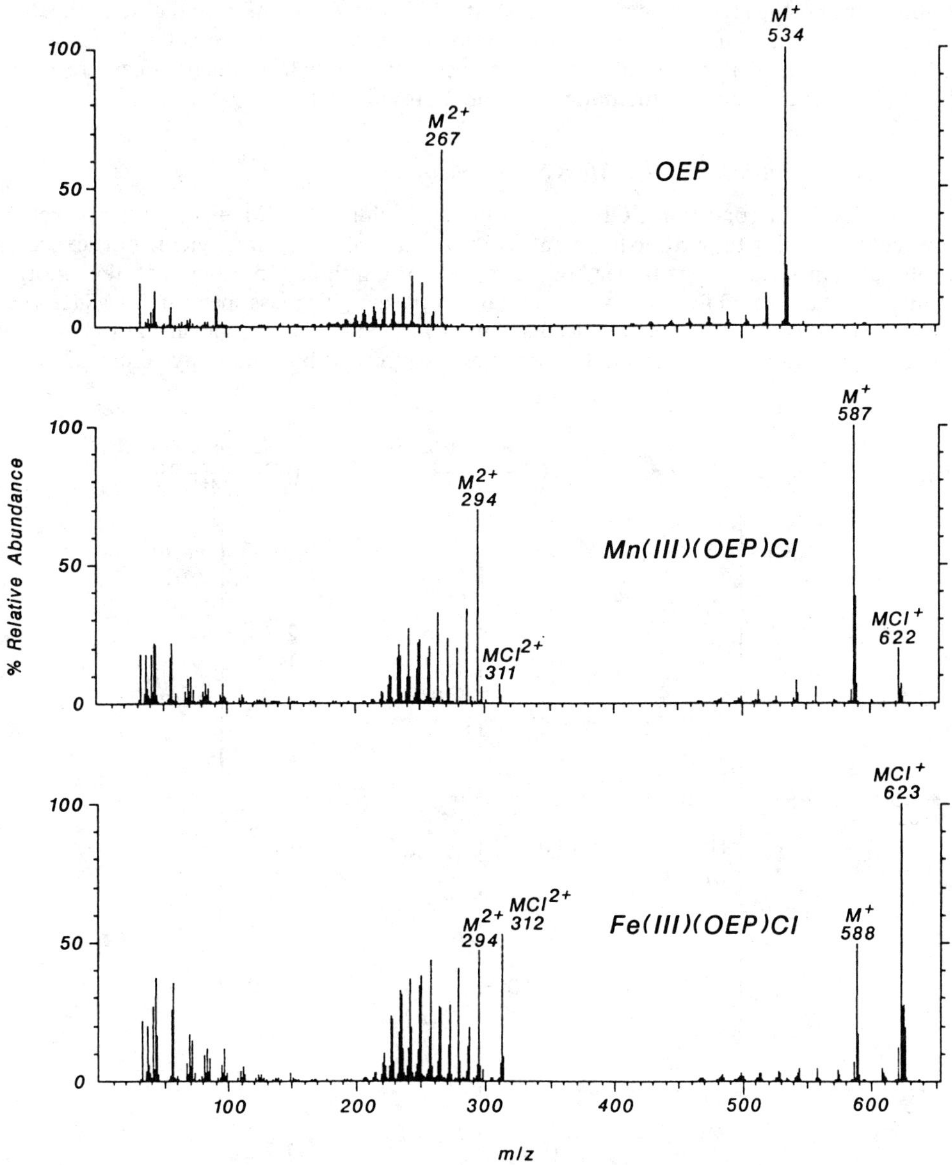

FIG. 3—*Electron ionization mass spectra of metalated OEP derivatives and of metal-free OEP.*

spectra of the OEP derivatives show the expected singly and doubly charged molecular ions, and abundant doubly charged fragment ions (Fig. 3). The three trivalent metal complexes showed both the molecular ion including the chloride ligand, and the ligand-free molecular ion. The doubly charged β-cleavage fragments in each sample were much more intense than the singly charged β-cleavage fragments. The metalated derivatives all showed more β-cleavage fragmentation than the metal-free OEP did. In fact, the greater the stability index of the metalloporphyrin, the greater the relative abundance of doubly charged ions compared to that of the singly charged molecular ion (Fig. 4). Also, the higher the stability index, the greater the abundance of doubly charged fragments relative to the doubly charged molecular ion. These observations can be attributed to greater charge localization on the metal atom, withdrawing more electron density from the periphery, resulting in enhanced β-cleavage fragmentation, as noted previously [*3*].

Ammonia Chemical Ionization Mass Spectrometry of Porphyrins

Under NH_3 CI conditions, OEP is protonated, yielding the $(M + H)^+$ ion, as well as ions corresponding to pyrrolic fragments (Fig. 5). Each of the samples yielded the expected protonated molecular ions under NH_3 CI conditions, although no ions with the chloride ligand intact appeared for the trivalent metal complexes. The mass spectrum of Mg(II)OEP (Fig. 6) showed the $(M + H)^+$ molecular ion at m/z 557, as well as a substantial amount of ammonia adduct at m/z 574. The samples also formed both monopyrrolic and dipyr-

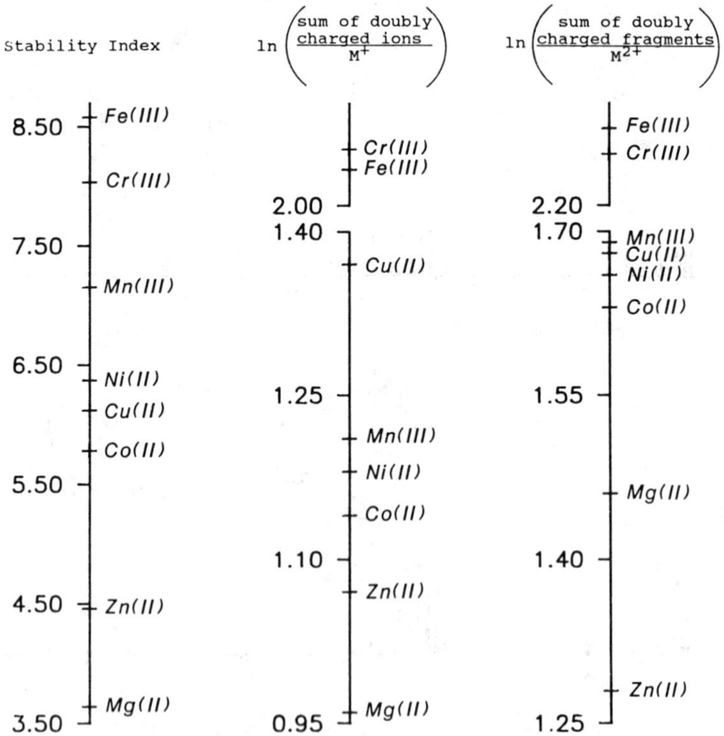

FIG. 4—*Comparison of stability index with extent of doubly charged β-cleavage fragment formation in EI mass spectra.*

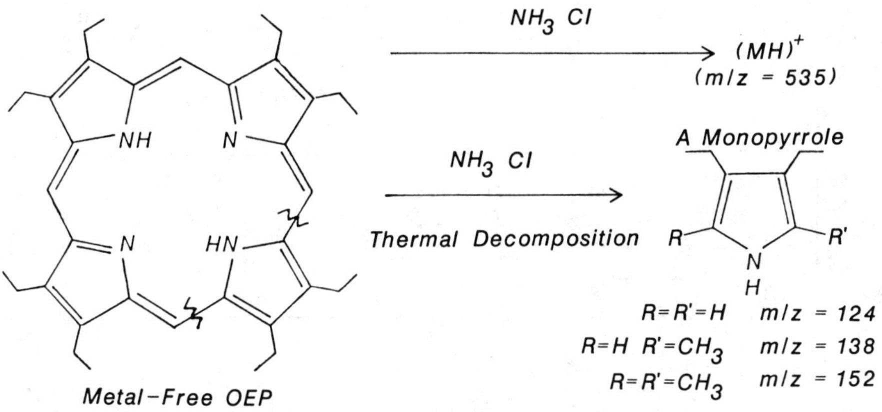

FIG. 5—*Fragmentation pathways of metal-free OEP under NH_3 CI conditions.*

rolic fragment ions together with a variety of NH_3 pyrrole adducts. Three of the metalloporphyrins, FE(III)OEP-Cl, MN(III)OEP-Cl, and CU(II)OEP, are likely to undergo reduction in the ammonia plasma, and are listed in their reduced forms.

In reducing solutions, Cu(II) porphyrins have been observed to be reduced to Cu(I) [*9*]. Under NH_3 desorption CI conditions, Cu(II) porphyrins have been observed to completely demetalate and fragment as metal-free porphyrins [*6*]. This is due to reduction of Cu(II) to Cu(I), which is readily extruded. If Cu(II) porphyrins are reduced under NH_3 CI conditions, it is likely that Fe(III) and Mn(III) porphyrins are also reduced in the NH_3 plasma, as these metal ions have higher standard reduction potentials than Cu(II) in solution [*10*]

$$\mathrm{Cu(II)} + e^- \rightleftarrows \mathrm{Cu(I)} \qquad E^\circ = +\,0.158\ V$$
$$\mathrm{Fe(III)} + e^- \rightleftarrows \mathrm{Fe(II)} \qquad E^\circ = +\,0.770\ V$$
$$\mathrm{Mn(III)} + e^- \rightleftarrows \mathrm{Mn(II)} \qquad E^\circ = +\,1.51\ V$$

Earlier work by Djerassi and others, involving NH_3 desorption CI of porphyrins, resulted in porphyrinogen $[(M + 6H + H)^+]$ formation [*6*]. It was postulated that the porphyrinogen ion was produced by repeated attacks of NH_4^+ on the porphyrin, and that it was the less-stable porphyrinogen that fragmented into the observed mono-, di-, and tripyrrolic components. In the present study in which a solids probe was used, neither the porphyrinogens nor tripyrroles were seen. If the pyrrole fragment ions were formed by fragmentation of the porphyrin $(M + H)^+$ ion in the gas phase, the intensity of the pyrrolic ions should increase and decrease with that of the molecular ion [*7*]. In this current work, the mass chromatogram for the pyrrolic ions does not track that of the molecular ion (Fig. 7). Thus it is likely that the monopyrroles and the limited amounts of dipyrroles observed probably arise from a process that is different from that which Djerassi described for NH_3 desorption CI. In this case, it appears that the pyrroles arise out of thermal decomposition of the porphyrin in the ammonia plasma as it resides on or evaporates off of the heated solids probe vial surface.

The extent of pyrrole formation is very dependent on the specific metal in the metalloporphyrin. The greater the stability index, the lower the relative abundance of pyrroles formed, compared to $(M + H)^+$ (Fig. 8). While Cu(II)(OEP) and Ni(II)(OEP) have very similar stability indices, Cu(II)(OEP) yields significantly higher intensities of pyrroles than does Ni(II)(OEP). In fact, Cu(II)(OEP) behaves very much like metal-free OEP. This sup-

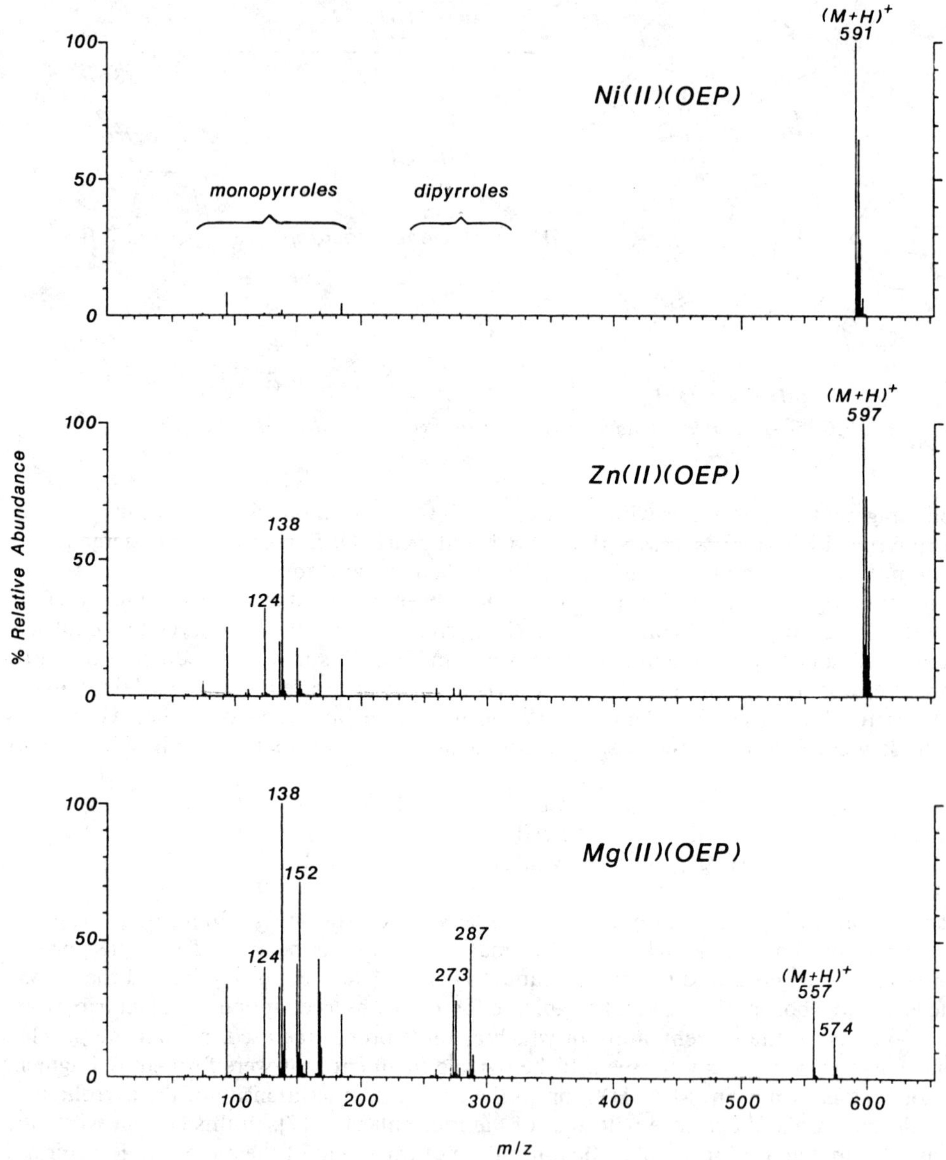

FIG. 6—*Ammonia chemical ionization mass spectra of metalated OEP derivatives and of metal-free OEP.*

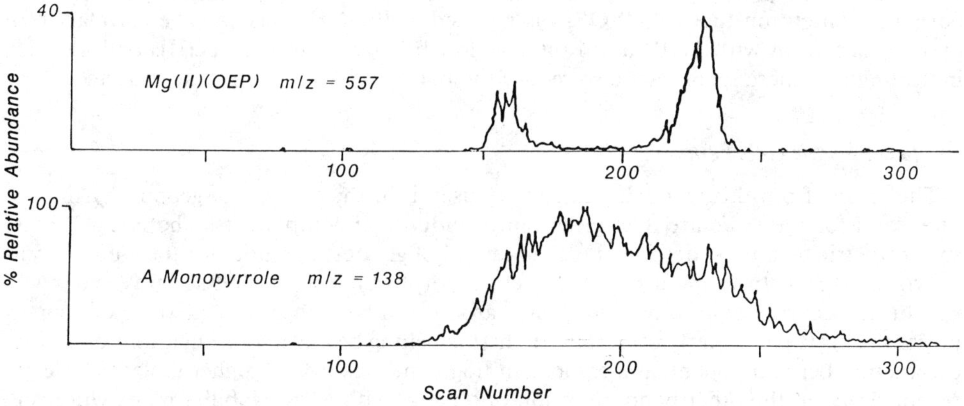

FIG. 7—*Typical NH_3 CI mass chromatograms of a protonated monopyrrole and the protonated molecular ion Mg(II)OEP.*

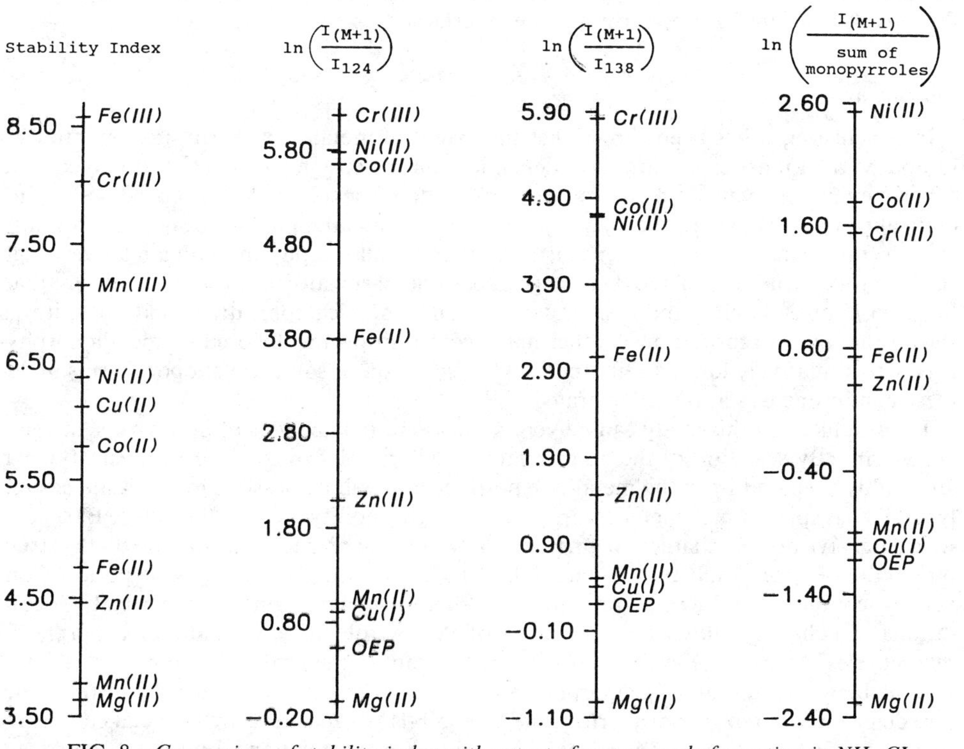

FIG. 8—*Comparison of stability index with extent of monopyrrole formation in NH_3 CI mass spectra.*

ports the contention that Cu(II)(OEP) is reduced in the NH_3 plasma. The correlation of pyrrole formation with Fe(II) and Mn(II) is much better than with Fe(III) and Mn(III), indicating that these metals are also reduced under NH_3 CI conditions, as expected.

Implications for Geochemistry

The results from this work should be very useful for the analysis of geoporphyrin mixtures, and for the structure determination of individual components. The NH_3 CI mass spectrometric technique described is the method of greatest potential for the study of porphyrin mixtures. Inserting a metal to yield a porphyrin with a high stability index will permit an accurate determination of the carbon number range of a mixture of geoporphyrins. This is particularly important in distinguishing between molecular ions of a lower carbon number component, and coincident fragment ions from a higher molecular weight component. On the other hand, selection of a metal with a low stability index will result in the generation of pyrrolic fragment ions, whose distributions may be used as fingerprints for the comparison of one mixture with another.

Both EI and NH_3 CI should be of value for the structure determination of individual geoporphyrins. A metal complex with a high stability index may be selected to better determine the types of substituent around the geoporphyrin periphery by EI mass spectrometry. Conversely, a metal complex with a low stability may be used for NH_3 CI mass spectrometry. By careful study of the monopyrrolic fragments, the identities of the substituents on the individual pyrrole rings of the porphyrin macrocycle may be deduced. If the dipyrrolic components are sufficiently abundant, for example, in Mg(II)OEP (Fig. 6), the sequence of the pyrrole rings in the porphyrin may be ascertained.

Conclusions

In conclusion, it has been shown that the extent of metalloporphyrin fragmentation in EI and NH_3 CI mass spectrometry is dependent on the specific metal atom involved. The experimental observations made concerning this dependence can be correlated with a theoretical physical constant, the stability index, of the metal complex. Under electron ionization conditions, it has been demonstrated that metalloporphyrins with a relatively high stability index tend to undergo β-cleavage of peripheral substituents to a greater extent than those with lower stability indices. Under ammonia chemical ionization conditions, it was shown that pyrrole formation via thermal decomposition was favored by metalloporphyrins with a relatively low stability index. The reduction of some metalloporphyrins under these conditions was also demonstrated.

The results from this study can be very valuable in the analyses of porphyrins and geoporphyrins. By selection of the appropriate metalloporphyrin complexes, most of which are readily prepared from the metal-free porphyrin, it will be possible to select the desired type of mass spectrum. A metal complex with a high stability index will produce EI spectra with extensive doubly charged fragmentation, which will provide information on the structure of the peripheral substituents, and NH_3 CI spectra with very little fragmentation. Conversely, complexes of low stability index will yield pyrrolic fragments under NH_3 CI conditions, which may allow the sequencing of the pyrrole rings on individual porphyrin macrocycles. All of these techniques will prove useful in the study of geoporphyrins. They will facilitate correlation of geoporphyrin mixtures from different sources, and the structure determination of geoporphyrins. The EI and NH_3 CI mass spectra of metal complexes that have not been discussed here may be predicted by correlation with the calculated stability index of the complex. Finally, it was also noted in this study that some metalloporphyrins are reduced in the NH_3 CI plasma, and that this is important to know if one is to correctly predict the extent of fragmentation of these compounds. The criteria given for reduction to occur, which were based on solution chemistry, allows predictions to be made as to which metals will be reduced and which will not be.

References

[1] Filby, R. H. and van Berkel, G. J. in *Metal Complexes in Fossil Fuels: Geochemistry, Characterization, and Processing,* ACS Symposium Series 344, R. H. Filby and J. F. Branthaver, Eds., American Chemical Society, Washington, DC, 1987, pp. 2–39.

[2] Baker, E. W. and Palmer, S. E. in *The Porphyrins, Vol. 1, Structure and Synthesis Part A,* D. Dolphin, Ed., Academic Press Inc., New York, 1978, pp. 485–551.

[3] Smith, K. M. in *Porphyrins and Metalloporphyrins,* K. M. Smith, Ed., Elsevier Scientific Publishing Company, Amsterdam, 1975, pp. 381–398.

[4] Budzikiewicz, H. in *The Porphyrins, Vol. 3, Physical Chemistry Part A,* D. Dolphin, Ed., Academic Press Inc., New York, 1978, pp. 395–461.

[5] Quirke, J. M. E. in *Metal Complexes in Fossil Fuels: Geochemistry, Characterization, and Processing,* ACS Symposium Series 344, R. H. Filby and J. F. Branthaver, Eds., American Chemical Society, Washington, DC, 1987, pp. 308–331.

[6] Tolf, B., Jiang, X., Wegmann-Szente, A., Kehres, L. A., Bunnenberg, E., and Djerassi, C., *Journal of the American Chemical Society,* Vol. 108, No. 7, April 1986, pp. 1363–1374.

[7] Kurlansik, L., Williams, T. J., Strong, J. M., Anderson, L. W., and Campana, J. E., *Biomedical Mass Spectrometry,* Vol. 11, No. 9, Sept. 1984, pp. 475–481.

[8] Johnson, J. V., Britton, E. D., Yost, R. A., Quirke, J. M. E., and Cuesta, L. L., *Analytical Chemistry,* Vol. 58, No. 7, June 1986, pp. 1325–1329.

[9] Buchler, J. W. in *Porphyrins and Metalloporphyrins,* K. M. Smith, Ed., Elsevier Scientific Publishing Company, Amsterdam, 1975, pp. 157–231.

[10] *CRC Handbook of Chemistry and Physics,* R. C. Weast, Ed., CRC Press, Inc., Boca Raton, 1979, pp. D155–D156.

Emilio J. Gallegos[1]

Gas Chromatography/Mass Spectrometry Analysis of *n*-Alkylbenzothiophenes Identified in Arabian Heavy Asphaltenes, Pyrolysis Oil

REFERENCE: Gallegos, E. J., **"Gas Chromatography/Mass Spectrometry Analysis of *n*-Alkylbenzothiophenes Identified in Arabian Heavy Asphaltenes Pyrolysis Oil,"** *Novel Techniques in Fossil Fuel Mass Spectrometry, ASTM STP 1019,* T. R. Ashe and K. V. Wood, Eds., American Society for Testing and Materials, Philadelphia, 1989, pp. 94–115.

ABSTRACT: Present concepts hold that the genesis of organic sulfur compounds is due mainly to secondary processes. The secondary process is a chemical reaction of oil with sulfur or sulfur species to give nonthiophenic sulfides and thiols first, then benzothiophenes, and ultimately stable dibenzothiophenes. This work describes the pyrolysis gas chromatography/mass spectrometry (GC/MS) identification of a homologous series of *n*-alkylbenzothiophenes extending from C_{10} to C_{30} in Arabian Heavy asphaltenes. Within each homolog there are at least 22 isomers. The mass spectra of these isomers typically show relatively intense molecular ions with base peaks at *m/z* 147, 161, or 175.

The striking similarity of the fragmentation profiles of the homologous benzothiophenes and homologous *n*-alkylbenzenes also present in the sample strongly suggests that the benzothiophenes are generated directly from the reaction of the alkylbenzenes with sulfur species. Reflux experiments with known alkylbenzenes and elemental sulfur produced an abundance of thiophene derivatives including benzothiophenes.

KEY WORDS: mass spectrometry, fossil fuels, pyrolysis, *n*-alkylbenzothiophenes

Present concepts hold that the genesis of organic sulfur compounds found in crudes is due mainly to secondary processes [*1*]. The secondary process is a chemical reaction of the oil with sulfur, that is, elemental sulfur, hydrogen sulfide, or perhaps some intermediate form of sulfur to give nonthiophenic sulfides and thiols first, then benzothiophenes, and ultimately stable dibenzothiophenes. In fact, it has been found that the ratio of benzothiophenes to dibenzothiophenes is a useful indicator of the maturity of crudes [*1*].

A homologous series of *n*-alkylbenzothiophenes extending from C_{10} to C_{30} has been identified in the pyrolysis oil of Arabian Heavy asphaltenes. Within each homolog there are at least 22 resolved isomers. The mass spectra of these isomers typically show relatively intense molecular ions with base peaks at *m/z* 147, 161, or 175. Genesis of the benzothiophenes is rationalized on the basis of sulfur reacting with *n*-alkylbenzenes that are also present in the Arabian Heavy pyrolysis oil.

The identification of *n*-alkylbenzothiophenes reported here for the first time and *n*-alkylbenzenes in the pyrolysis oil of Arabian Heavy asphaltenes supports the view of secondary introduction of sulfur in the formation of organic sulfur compounds in crudes. Further, this work suggests that there may be a direct formation of benzothiophenes from the reaction of sulfur species with the *n*-alkylbenzenes.

[1] Senior research associate, Chevron Research Company, Richmond, CA 94802–0627.

Experimental

Arabian Heavy asphaltenes were obtained by hot heptane (~80°C) precipitation from Arabian Heavy crude. Data were obtained both by on-line pyrolysis gas chromatography/mass spectrometry (GC/MS) of the asphaltenes and by GC/MS of a monoaromatic fraction and the remaining aromatics fraction from the asphaltene pyrolysis oil.

The pyrolysis GC/MS system has been described elsewhere [2]. The monoaromatics and remaining aromatics fraction were obtained as follows: the asphaltenes were pyrolyzed in a tube furnace under vacuum (<0.1 mm Hg) with a final temperature of 550°C for 1 h. The resulting pyrolysis oil was then separated into saturate, aromatic, and polar fractions by liquid chromatography (LC). The aromatic fraction was further separated using LC methods into a monochromatic fraction and a remaining aromatics fraction.

Both pyrolysis GC/MS and GC/MS data were obtained using a 60-m DB-1 fused silica capillary column from J & W Scientific in a 5700A Hewlett-Packard gas chromatograph. The column was coupled directly to the source of a VG ZAB or VG 7070H double focusing mass spectrometer. Both M/ΔM = 1000 and M/ΔM 10 000 resolution pyrolysis GC/MS and GC/MS runs were made on the ZAB instrument. The B^2/E pyrolysis GC/MSMS runs were made on the 7070H instrument. The B^2/E linked scanning is a technique for monitoring the unimolecular decomposition of precursor ions, M_1, to selected fragments, M_2, in the first field-free region of a normal geometry double focusing mass spectrometer [3,4] where

B = magnetic field
E = electrostatic field voltage

The accelerating voltage was held constant at 4000 V for these experiments.

The columns in all cases were programmed from −30 to 310°C at 2°C/min. Dedicated Finnigan Incos computer systems were used for instrument control and data acquisition. Data were collected at a repetition rate of 1.5 and 2.0 s from m/z 10 to 600.

Results and Discussion

The following data were obtained from the pyrolysis GC/MS analysis of Arabian Heavy asphaltenes at an MS resolution of ~1000.

Figure 1 shows the total ion reconstructed gas chromatogram resulting from the pyrolysis GC/MS of Arabian Heavy asphaltenes.

Figures 2 to 8 present a graphic summary of the major hydrocarbon and sulfur carbon types identified in the pyrolysate, including the homologous benzothiophenes that are the subject of this paper.

These figures show the even mass MAP display of the various homologous series arranged according to their Z category. These categories are defined by the general formula C_nH_{2n+z} and $C_nH_{2n+z}S$. The first member of the homologous series in each category is illustrated with a probable structure.

Close scrutiny of Fig. 6 reveals a striking pattern of peaks in the m/z 92, 106, 148, and 162 mass chromatograms. A magnified view of the pattern within the square in Fig. 6 is given in Fig. 9. The repeat pattern of a singlet at m/z 92, doublet at m/z 106, and multipet at m/z 120 has been reported previously [5]. These are thought to be due to sets of homologous series of n-alkylbenzenes each with a benzene ring substituent configuration that remains constant. The increase in molecular weight comes only from increasing the length of the n-alkyl side chain.

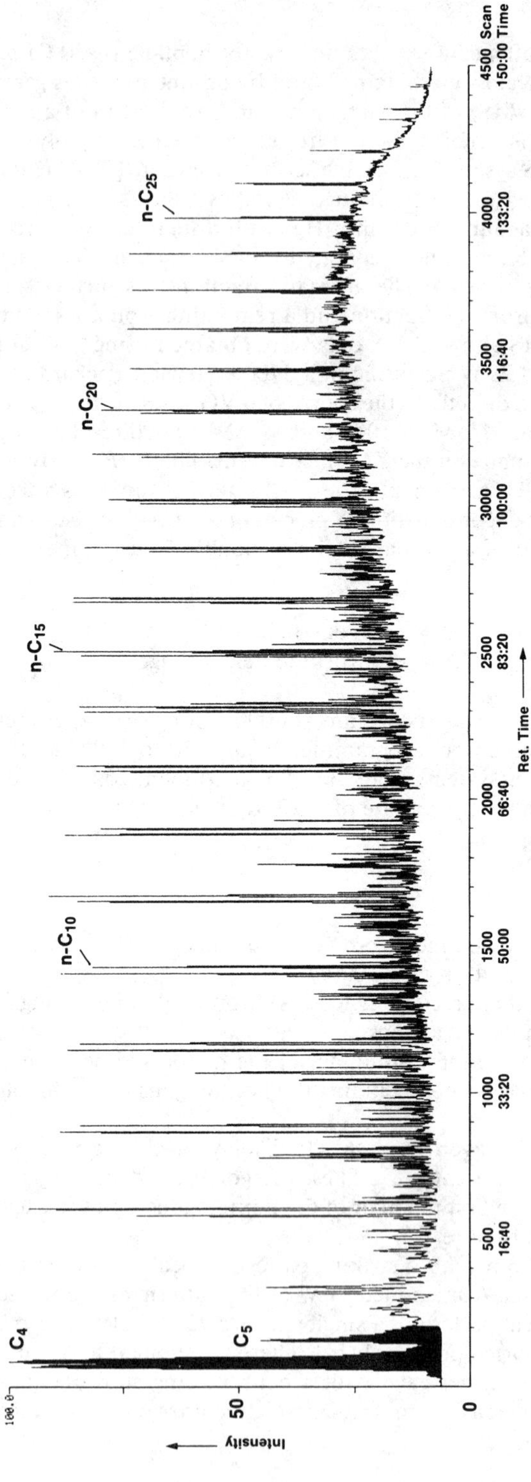

FIG. 1—*Reconstructed total ion gas chromatogram from pyrolysis GC/MS of Arabian Heavy asphaltenes.*

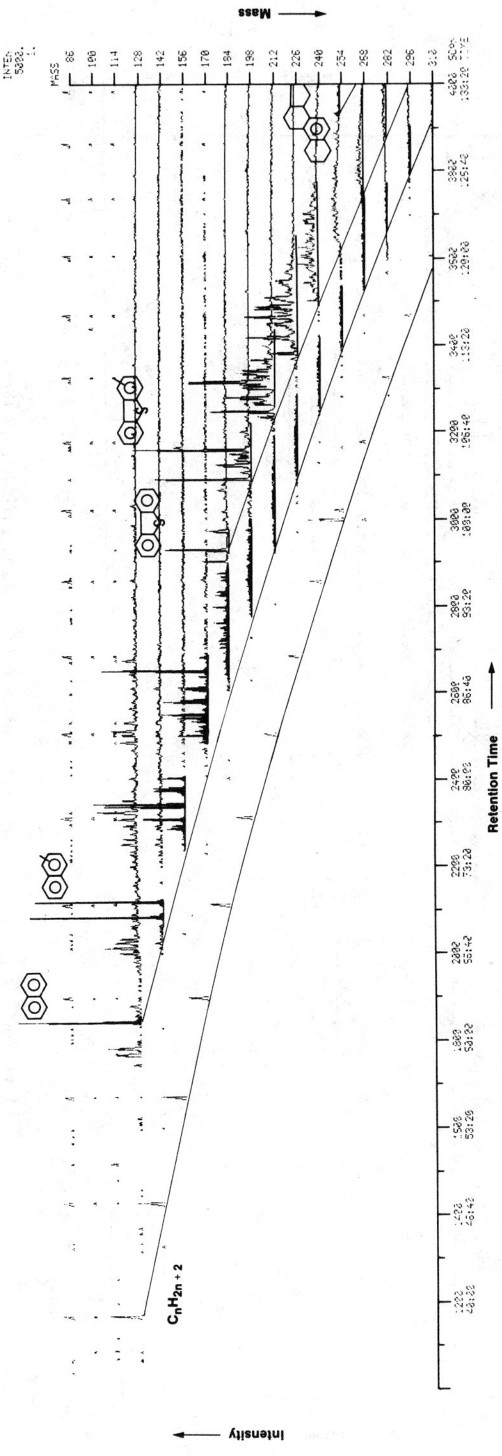

FIG. 2—$C_nH_{2n}{}^{+}{}_{2-2}$; $C_nH_{2n-16}S$ MAP from pyrolysis GC/MS of Arabian Heavy asphaltenes.

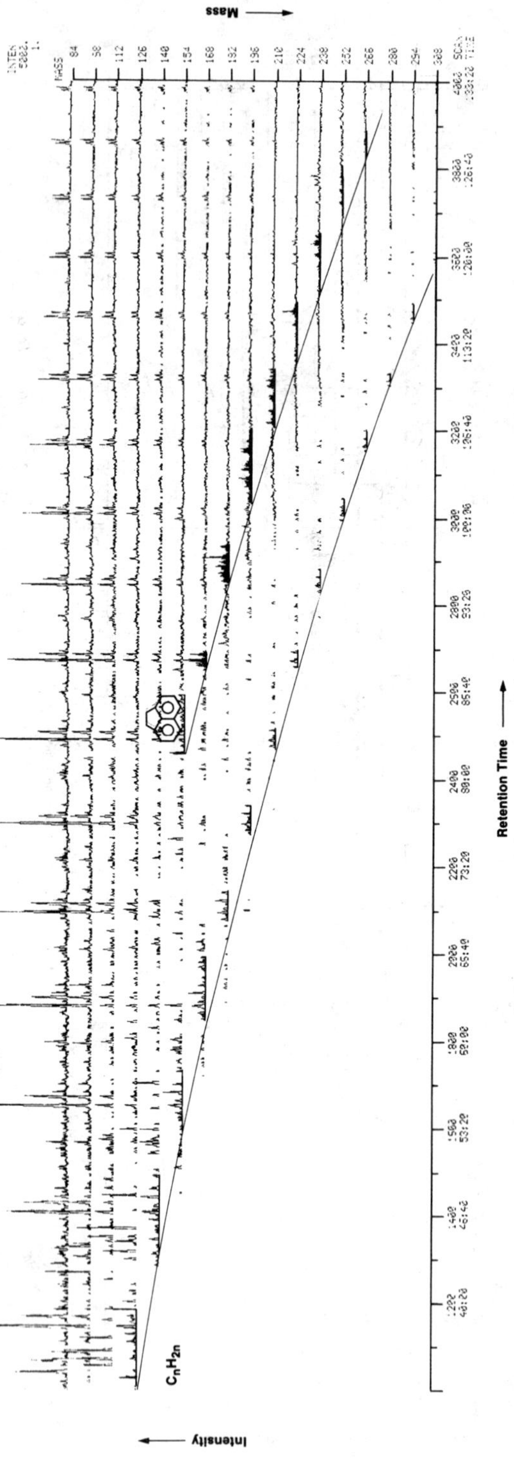

FIG. 3—$C_nH_{2n+0,-14}$ MAP from pyrolysis GC/MS of Arabian Heavy asphaltenes.

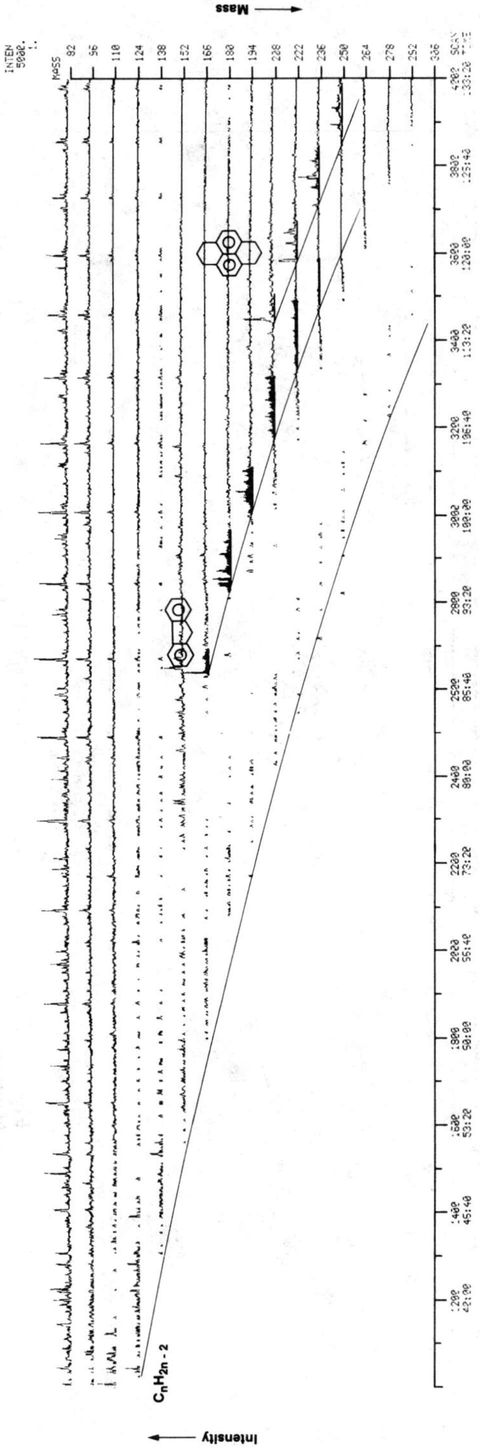

FIG. 4—$C_nH_{2n-2,-16}$ *MAP from pyrolysis GC/MS of Arabian Heavy asphaltenes.*

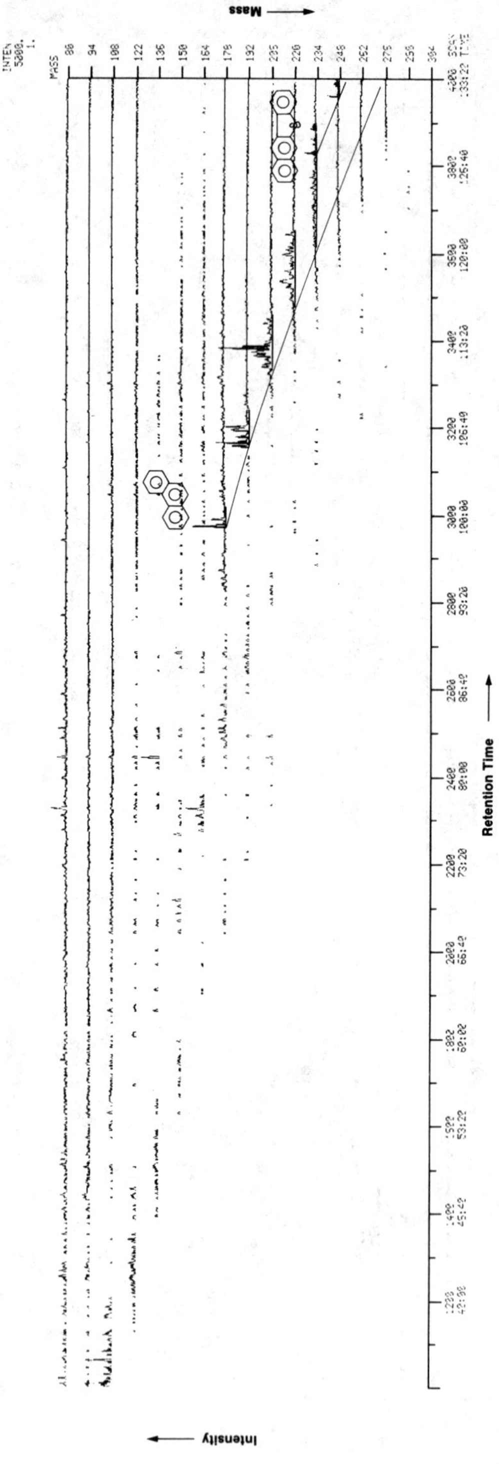

FIG. 5—$C_nH_{2n-4,-18}$; $C_nH_{2n-22}S$ MAP from pyrolysis GC/MS of Arabian Heavy asphaltenes.

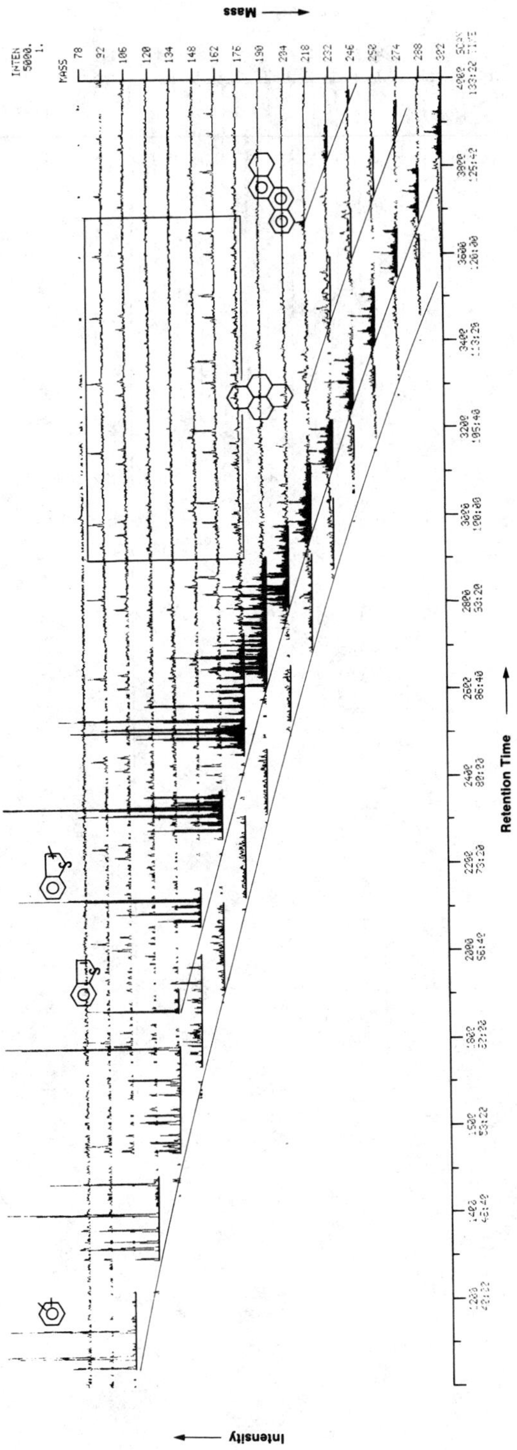

FIG. 6—$C_nH_{2n-6,-20}$; $C_nH_{2n-10}S$ MAP from pyrolysis GC/MS of Arabian Heavy asphaltenes.

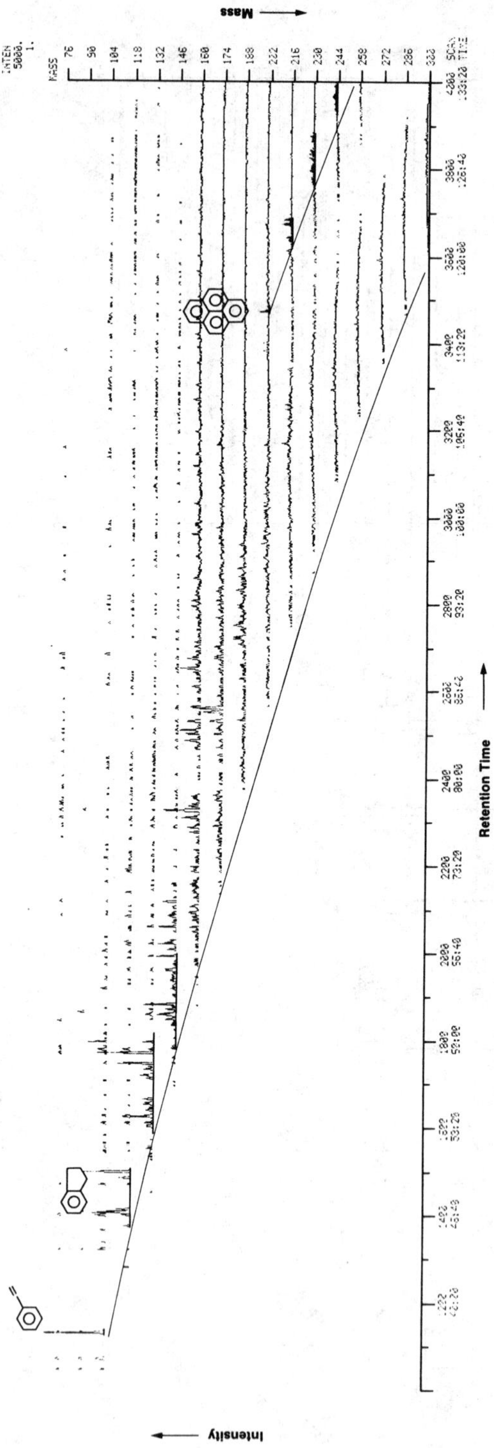

FIG. 7—$C_nH_{2n-8,-22}$ *MAP from pyrolysis GC/MS of Arabian Heavy asphaltenes.*

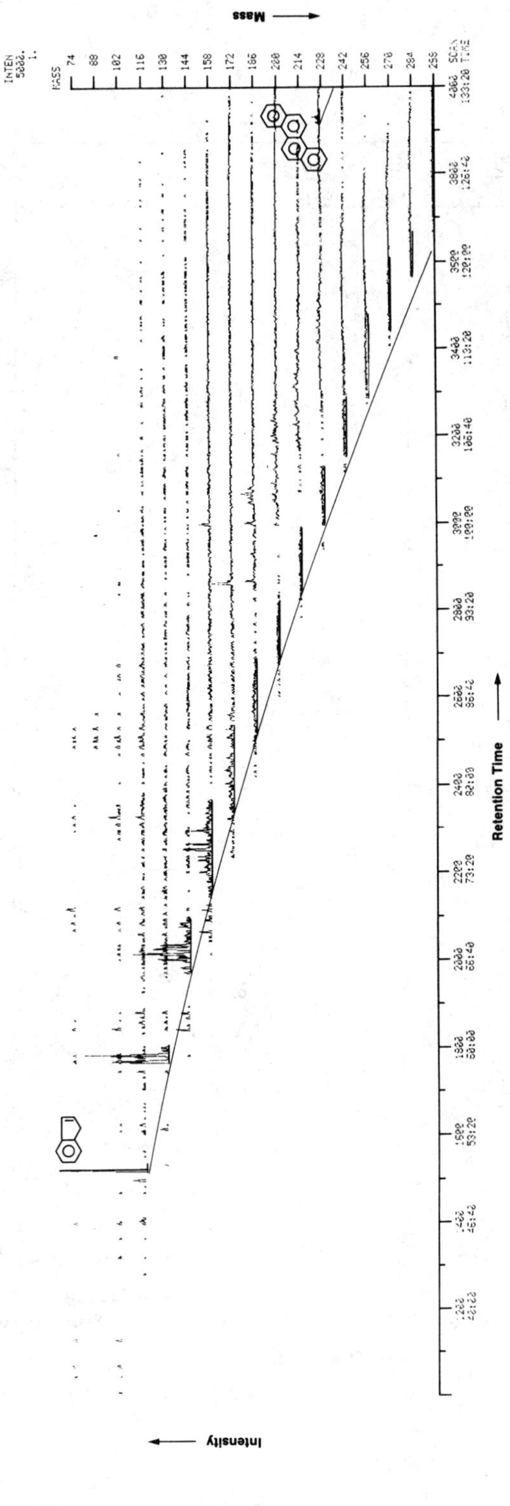

FIG. 8—$C_nH_{2n-10,-24}$ *MAP from pyrolysis GC/MS of Arabian Heavy asphaltenes.*

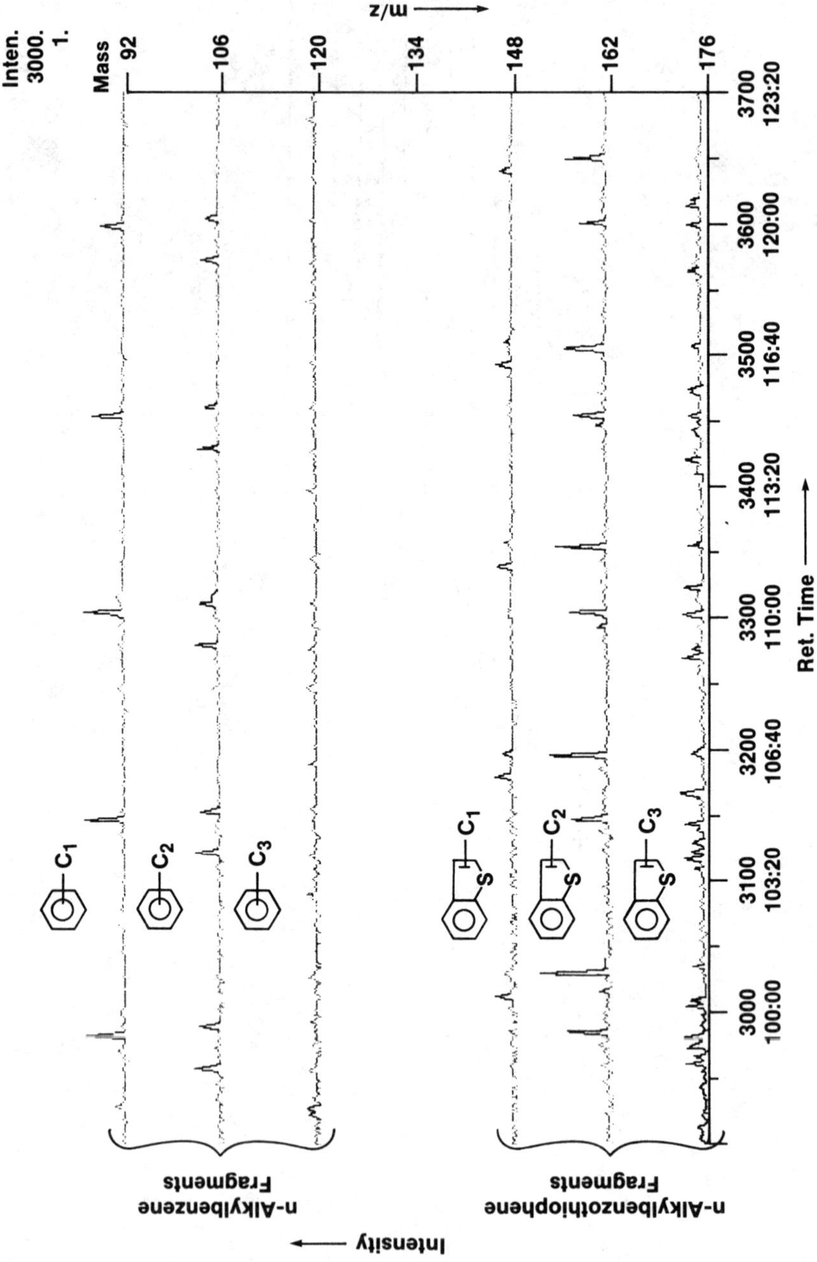

FIG. 9—*n-alkylbenzene and* n-*alkylbenzothiophenes fragment ion homologs. Magnified view from Fig. 6.*

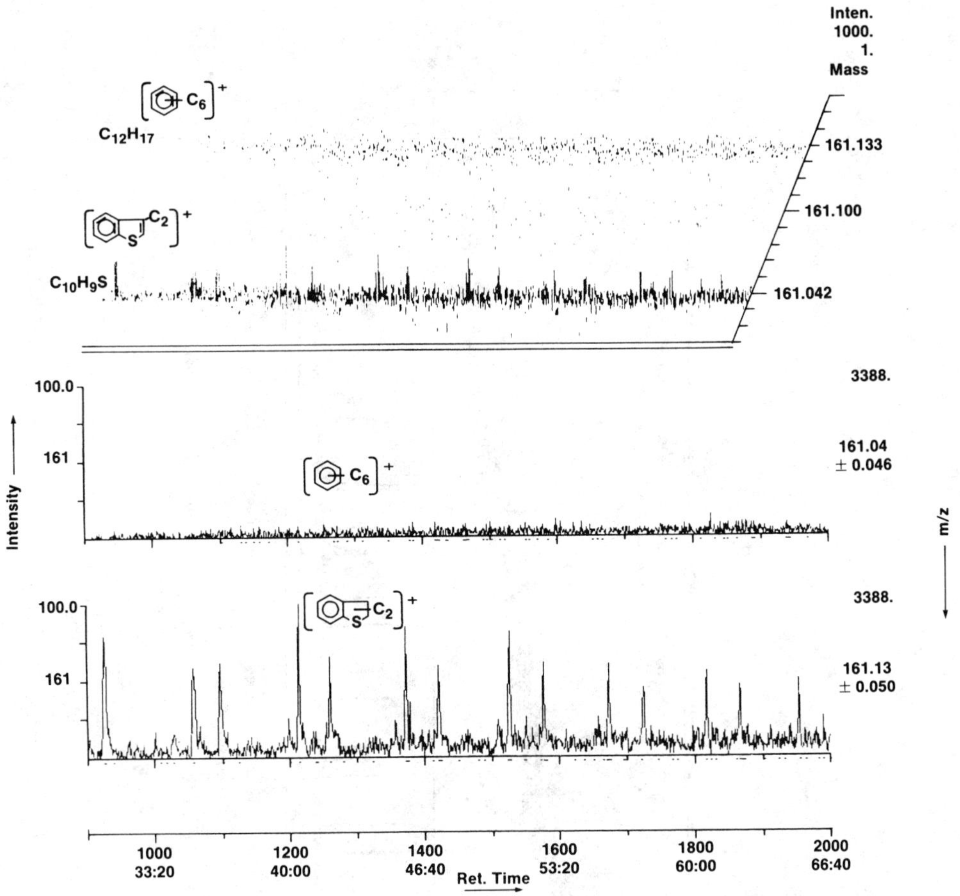

FIG. 10—*Accurate mass MAP and mass chromatogram of fragment* m/z *161 at M/ΔM = 10 000 from pyrolysis GC/MS of Arabian Heavy asphaltenes.*

This same reasoning may be applied to the singlet, doublet, and multiplet pattern of the "benzothiophene" peaks at *m/z* 148, 162, and 176. Individual mass spectra of these mass chromatographic peaks show that in fact *m/z* 147, 161, and 175 are the base peaks.

The presence of sulfur in Fragments 147, 161, and 175 must be confirmed since at nominal mass resolution benzothiophenes and alkylbenzenes are isobaric. Pyrolysis GC/MS at 10 000 mass resolution confirms the presence of sulfur as shown in Fig. 10. This figure shows the accurate mass MAP and mass chromatogram of fragment *m/z* 161 due to $C_{12}H_{17}$ and $C_{10}H_9S$. The same data shows that sulfur is also present in fragment ions at *m/z* 147 and 175.

Analysis by pyrolysis GC/MSMS, B^2/E scans, of the Arabian Heavy asphaltenes for the progenitors of *m/z* 161, $C_{10}H_9S$, provides the data shown in Fig. 11. This is a pyrolysis GC/MSMS MAP of precursors $C_{15}S$ through $C_{32}S$ to *m/z* 161. Background is due to C_{16} to C_{34} alkylbenzene precursors to *m/z* 161 $C_{12}H_{17}$. The M* is defined as follows

$$M^* = \frac{(M_2)^2}{M_1}$$

where M_2 is fragment ion *m/z* 161 and M_1 corresponds to the ion precursor to M_2.

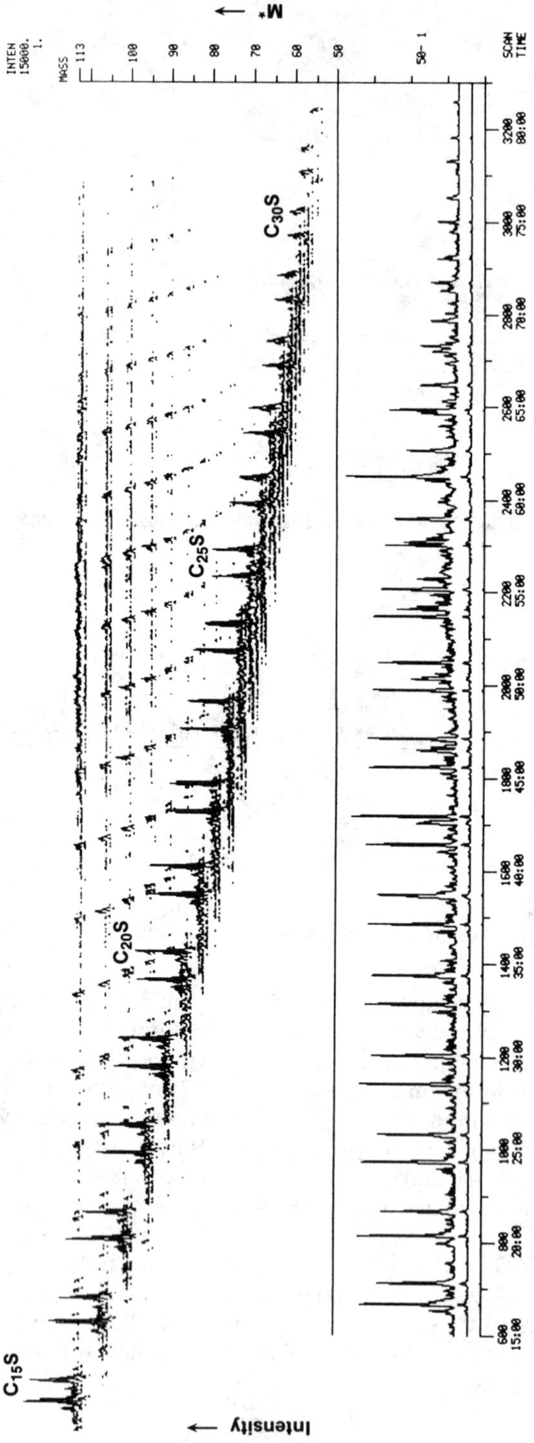

FIG. 11—*GC/MSMS MAP of* $C_{15}S$ *through* $C_{32}S$ *precursors to fragment ion m/z 161.*

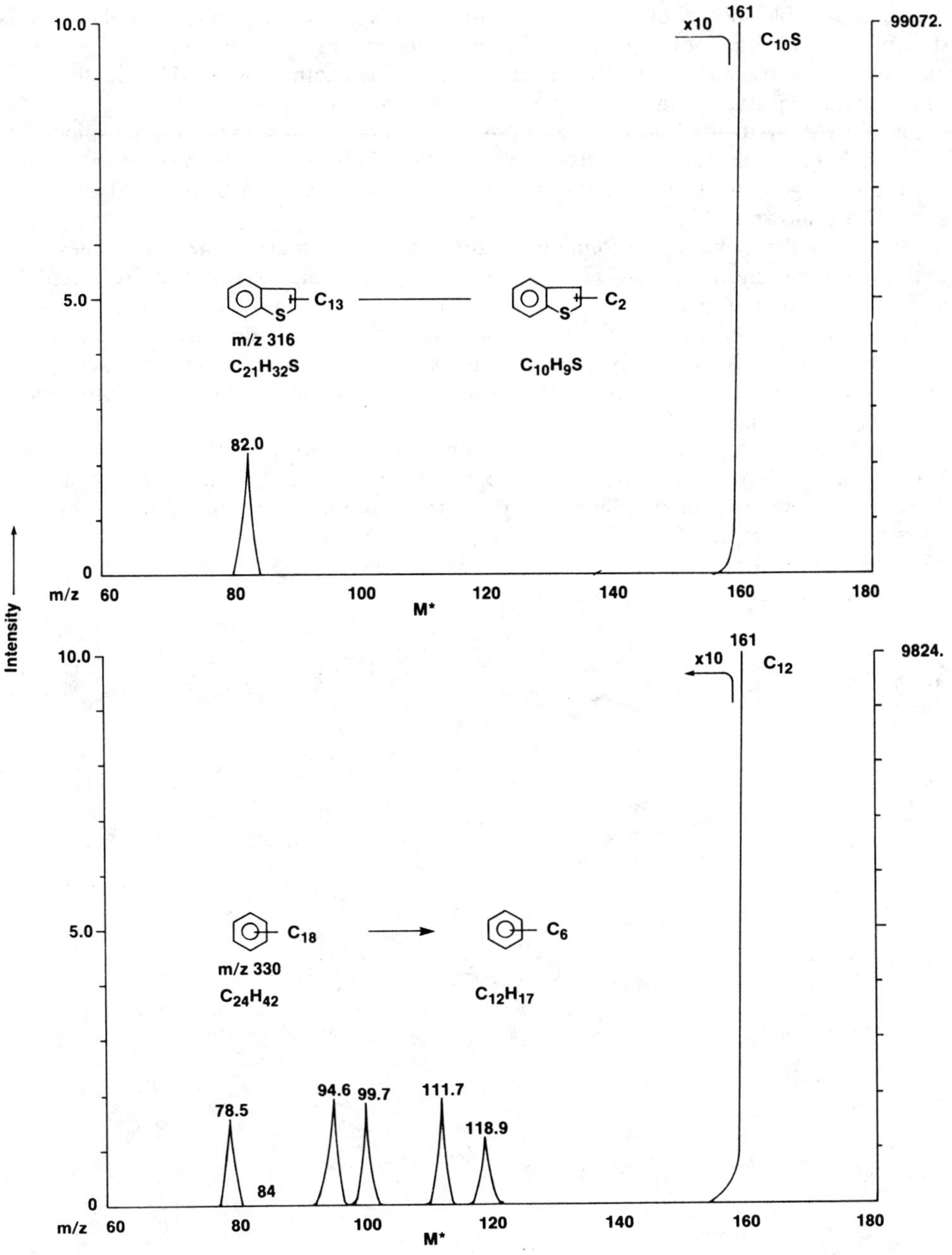

FIG. 12—*Metastable spectra of a* C_{21} n-*alkylbenzothiophene and an adjacent* C_{24} n-*alkylbenzene.*

The metastable spectra of the C_{21} benzothiophene and an adjacent C_{24} alkylbenzene is shown in Fig. 12. The molecular ion is the major precursor to fragment $C_{10}H_9S$ at m/z 161, indicating a simple side chain cleavage in the case of benzothiophenes. The alkylbenzene shows sequential fragmentation to give the $C_{12}H_{17}$ ion at m/z 161.

The GC/MS data on the monoaromatic fraction from Arabian Heavy asphaltenes reveals a homologous series of *n*-alkylbenzenes. The C_{14} through C_{17} homologs with major even mass fragments at m/z 92, 106, and 120 with molecular ions at m/z 190, 204, 218, and 232 are shown in Fig. 13.

Data from the remaining aromatic fraction reveal the *n*-alkylbenzothiophenes. The major fragment ions at m/z 147, 161, and 175 with molecular ions at m/z 218, 232, 246, and 260 are shown in Fig. 14.

Again note that the alkylbenzenes show a singlet at m/z 92, a doublet at m/z 106, and multiplets at m/z 120. Similarly, the *n*-alkylbenzothiophenes show a singlet at m/z 147, a doublet at m/z 161, and a multiplet at m/z 175. This same pattern for both sets of homologs repeats from $\sim C_{13}$ to $>C_{30}$.

This striking pattern and singlet, doublet, and multiplet for both the alkylbenzene and benzothiophene homologs strongly suggests that the *n*-alkylbenzothiophenes are generated directly from the reaction of the *n*-alkylbenzenes with sulfur during diagenesis. This is consistent with current understanding that thiophenes are a product of a secondary reaction of sulfur in some form with the oil. The following scheme is proposed to explain the genesis of the *n*-alkylbenzothiophene homologs

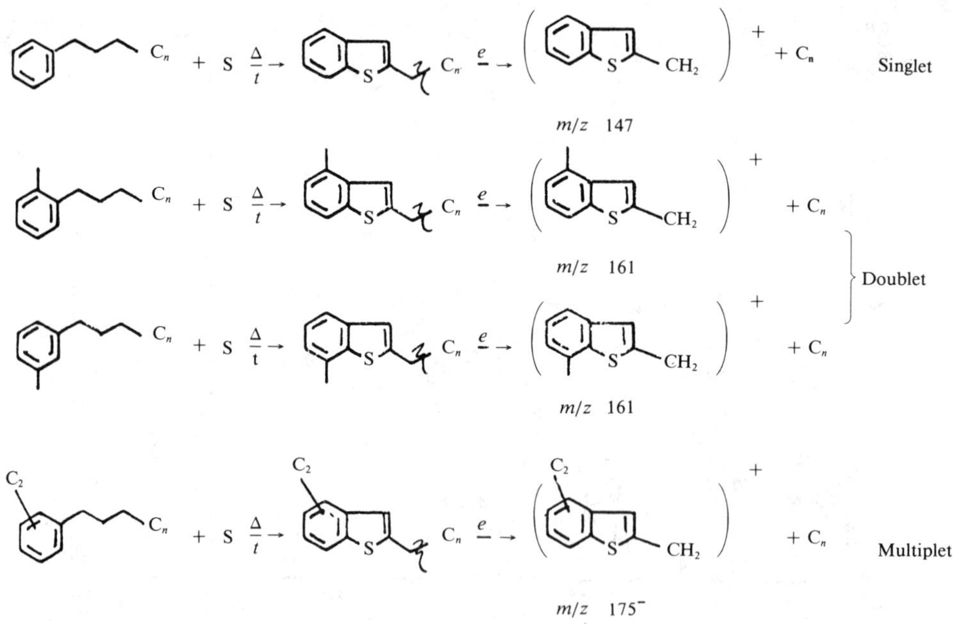

The ratio of concentration of *n*-alkylbenzenes to *n*-alkylbenzothiophenes is $\sim$0.6. The utter regularity of the retention times of the benzothiophene homologs from C_{13} to $>C_{30}$ is used as the basis for assuming that an increase in carbon number of the homologs is achieved by increasing the length of the *n*-alkyl side chain.

A close-up display of the isomers of the $C_{15}H_{20}S$ *n*-alkylbenzothiophene homolog is shown in Fig. 15. Shown are the m/z 147, 161, and 175 fragmentation mass chromato-

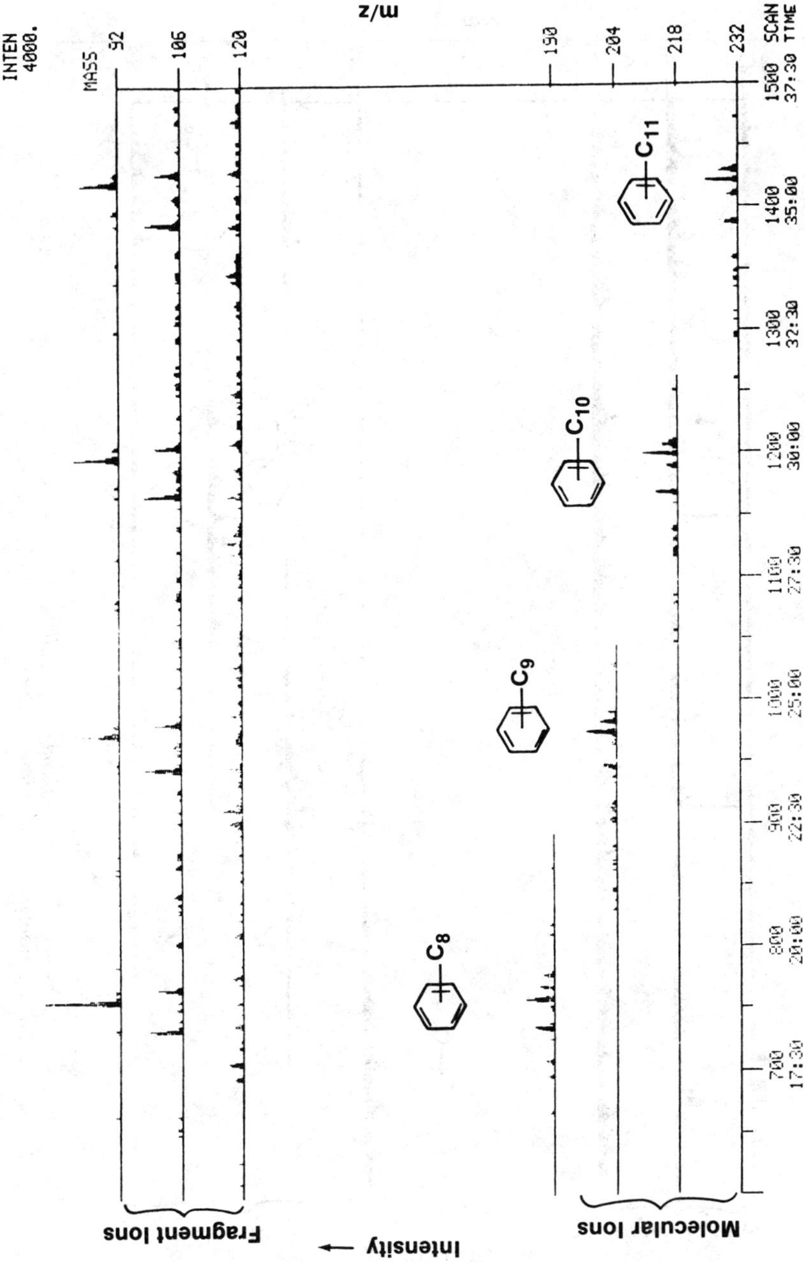

FIG. 13—*MAP of the C_{14} through C_{17} n-alkylbenzene homologs identified in the mono-aromatic fraction from Arabian Heavy asphaltenes pyrolysis oil.*

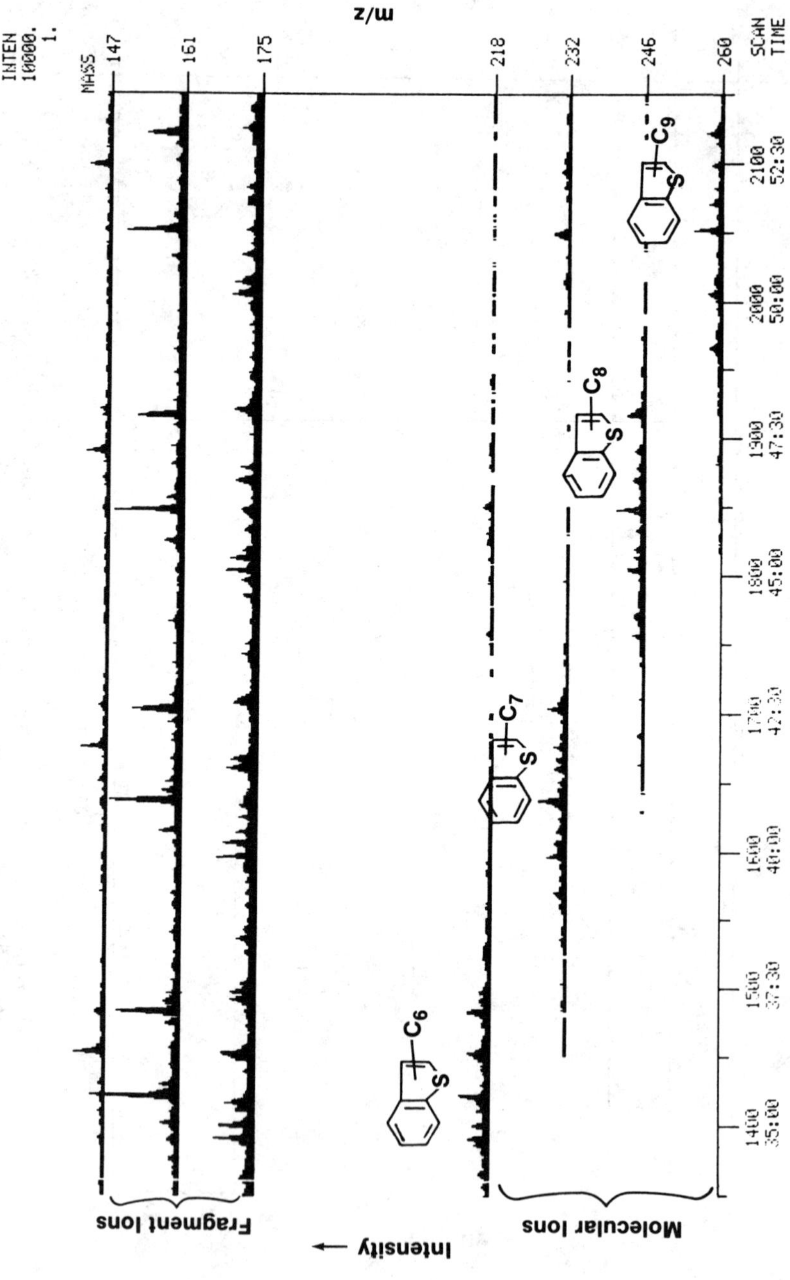

FIG. 14—*MAP of the* C_{14} *through* C_{17} n-*alkylbenzothiophene homologs identified in the remaining aromatics fraction from Arabian Heavy asphaltenes pyrolysis oil.*

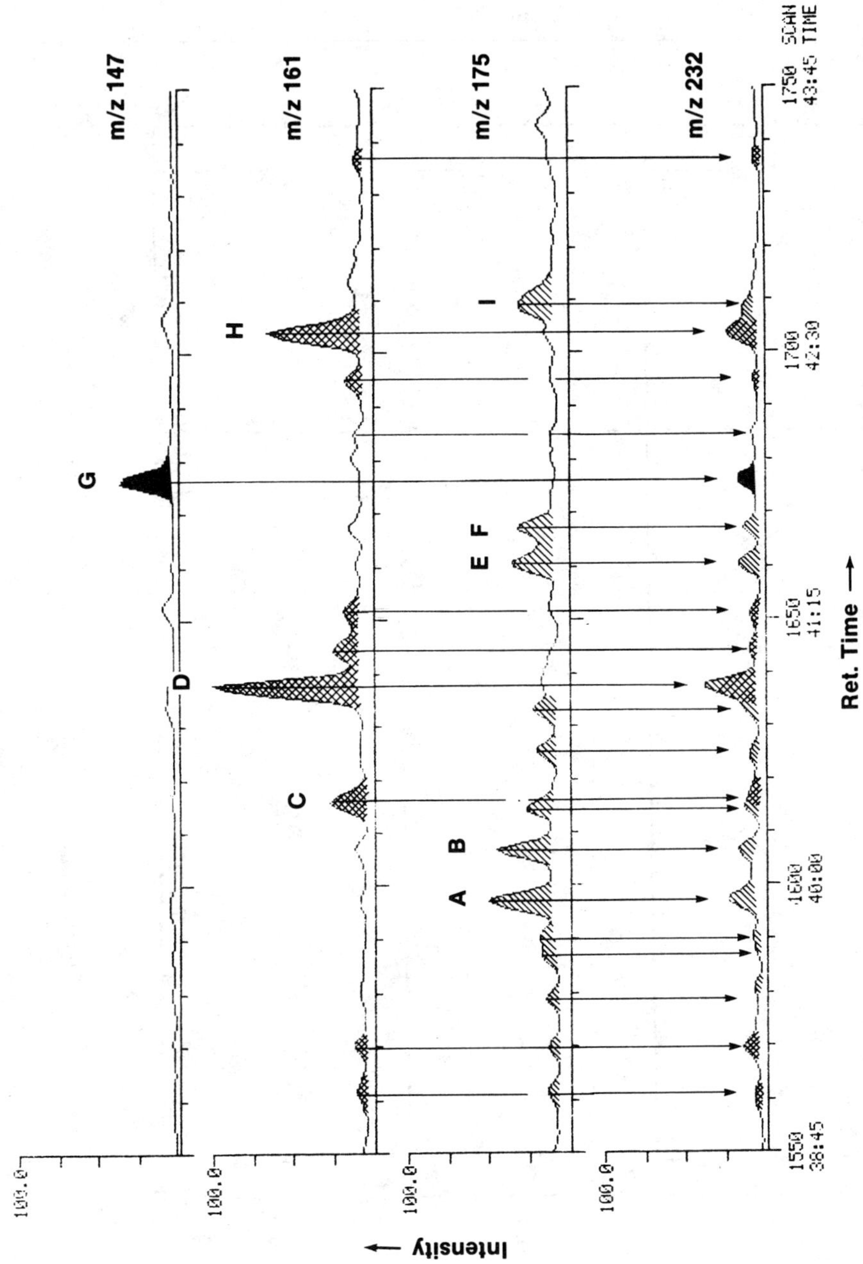

FIG. 15—*Mass chromatograms of the molecular and fragment ions due to the* $C_{15}H_{18}S$ n-*alkylbenzothiophene isomers identified in the remaining aromatics fraction from Arabian Heavy asphaltenes pyrolysis oil.*

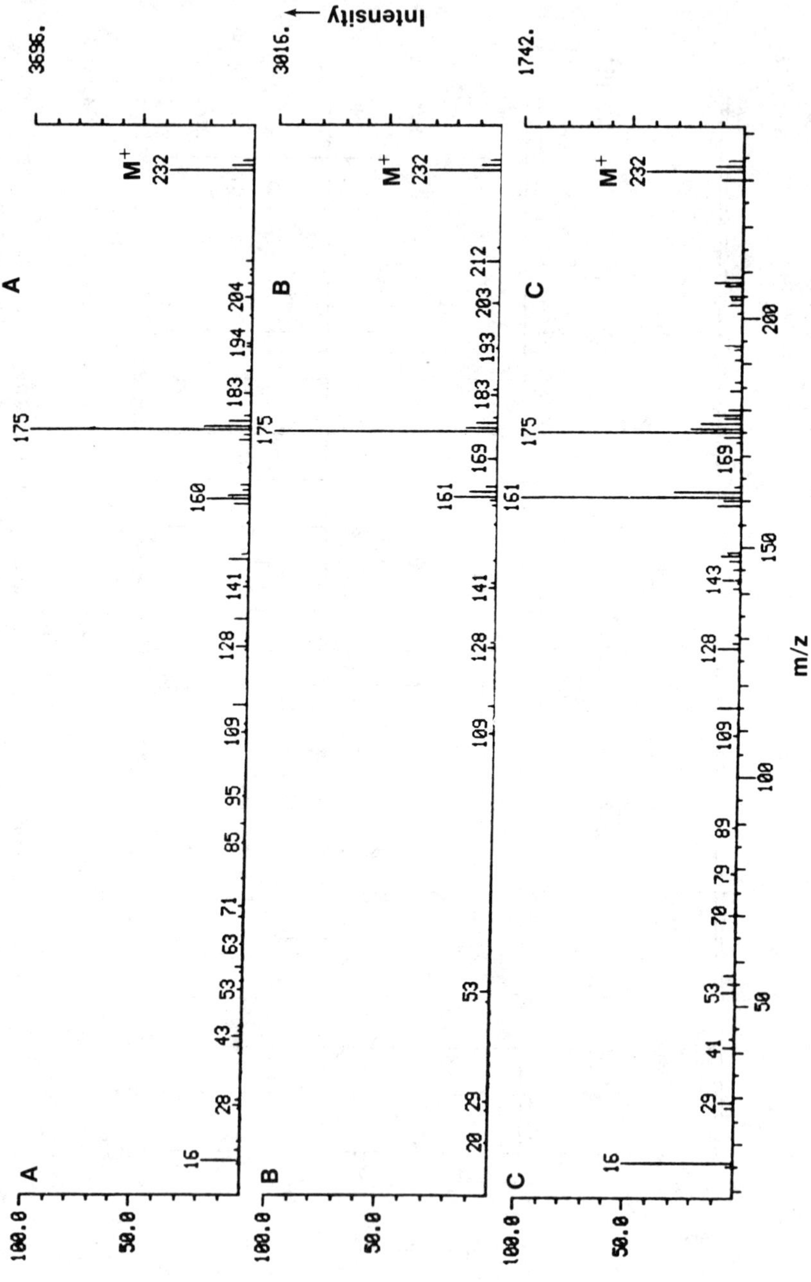

FIG. 16—Mass spectra of $C_{15}H_{18}S$ Isomers A, B, and C noted in Fig. 15.

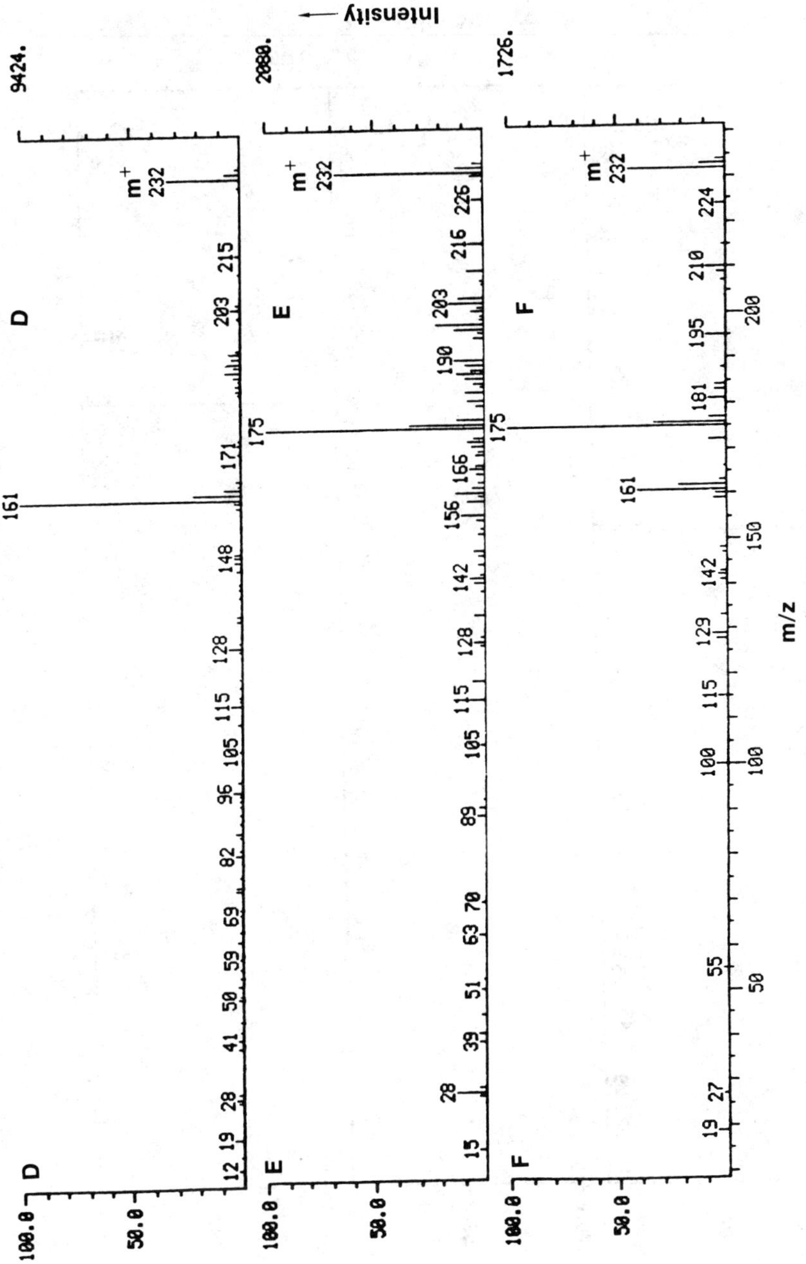

FIG. 17—*Mass spectra of $C_{15}H_{18}S$ Isomers D, E, and F noted in Fig. 15.*

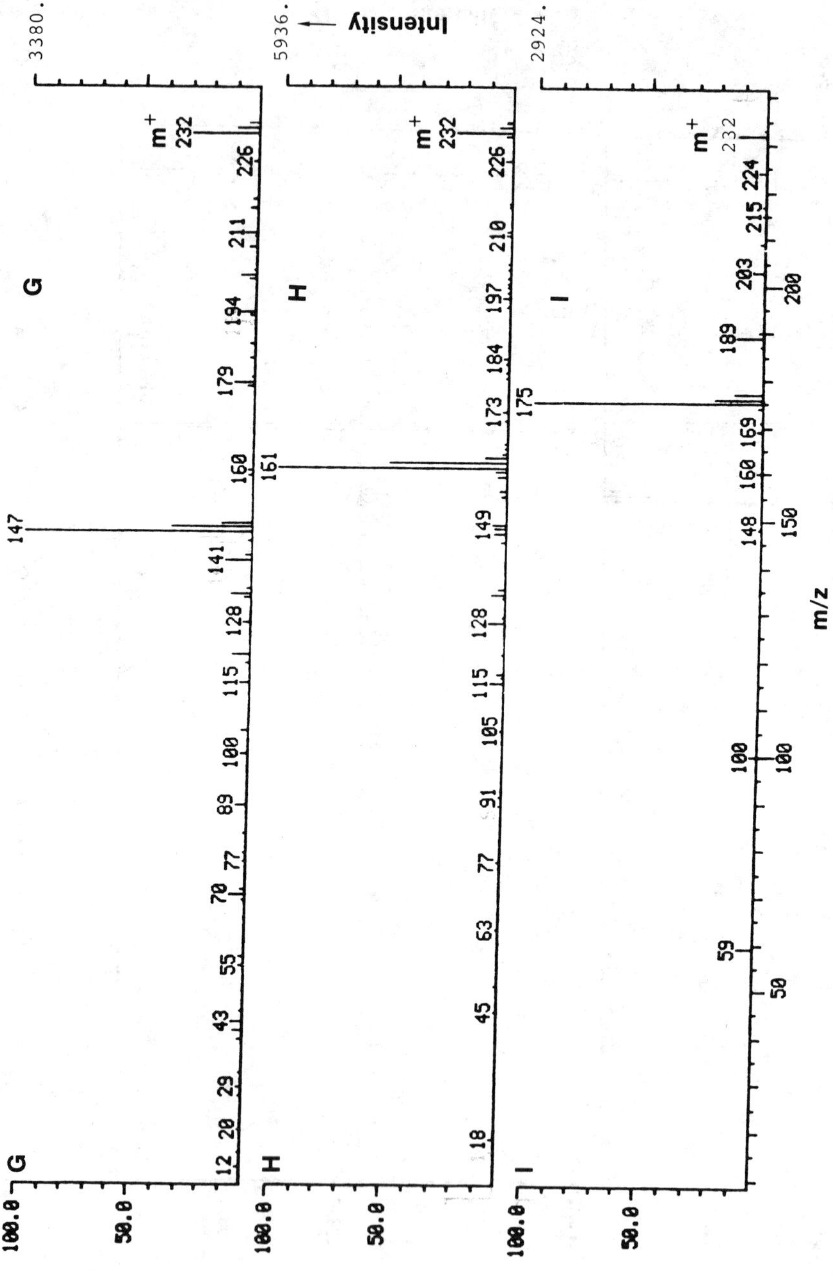

FIG. 18—*Mass spectra of $C_{15}H_{18}S$ Isomers G, H, and I noted in Fig. 15.*

grams and the molecular ion mass chromatogram of *m/z* 232. The intensities are normalized to the most intense ions at *m/z* 161. There are at least 22 separate isomers having molecular ions at *m/z* 232. This same pattern of molecular and fragment ions is repeated from approximately C_{13} to $>C_{30}$. The mass spectra of nine major isomers, A through I, with molecular ions at 232 are given in Figs. 16, 17, and 18.

These same homolog series of *n*-alkylbenzenes and *n*-alkylbenzothiophenes are also present in Arabian Heavy and Iranian Light crudes [*5*].

Reflux experiments at 150°C with 1-phenyltetradecane and small amounts of 1-phenylhexadecane and sulfur in the presence of V_2O_5 yielded detectable amounts of phenyl-*n*-decyl thiophene (molecular weight (MW) 300), *n*-dodecylbenzothiophenes (MW 302), phenyl-*n*-dodecylthiophenes (MW 328), benzothiophenyl-*n*-octylthiophene (MW 328), tetrahydrodibenzothiophene (MW 188), diphenylthiophene (MW 236), nonylthiophene (MW 210), and many other as yet unidentified thiophene derivatives [*6,7*].

Conclusions

Multiple isomer homologs of *n*-alkylbenzothiophenes and *n*-alkylbenzenes have been identified in Arabian Heavy asphaltenes pyrolysis oil and in the Arabian Heavy crude itself.

The close similarity between the molecular ion and fragment ion mass chromatograms of the *n*-alkylbenzenes and *n*-alkylbenzothiophenes suggests that the latter is formed from *n*-alkylbenzenes through reaction with sulfur in some form. Further, this is a secondary reaction during maturation of the oil when contact with sulfur is made.

These data support the concept of secondary reaction of sulfur with alkylbenzenes to form benzothiophenes directly during maturation.

Acknowledgment

The author would like to thank Ms. L. A. Wraxall for the pyrolysis of the asphaltenes, Mr. F. Fago for the high-performance liquid chromatography aromatic and saturate isolation of Arabian Heavy asphaltenes pyrolyzate, and Dr. J. C. Fetzer for the monoaromatics separations.

References

[*1*] Gallegos, E. J., "Terpane-Sterane Release From Kerogen by Pyrolysis Gas Chromatography-Mass Spectrometry," *Analytical Chemistry,* Vol. 49, No. 9, Aug. 1975, pp. 1524–1528.

[*2*] Gallegos, E. J., "Analysis of Organic Mixtures Using Metastable Transition Spectra," *Analytical Chemistry,* Vol. 48, No. 9, Aug. 1976, pp. 1348–1351.

[*3*] Haddon, W. F., "Computerized Mass Spectrometer Linked Scan System for Recording Metastable Ions," *Analytical Chemistry,* Vol. 51, No. 7, June 1979, pp. 983–988.

[*4*] Gallegos, E. J., "Alkylbenzenes Derived From Carotenes in Coals by GC/MS," *Journal of Chromatographic Science,* Vol. 19, April 1981, pp. 177–182.

[*5*] Gallegos, E. J., unpublished results.

[*6*] White, C. M., Douglas, L. J., Schmidt, C. E., and Hacke, H. M., "Formation of Polycyclic Thiophenes From Reaction of Selected Polycyclic Aromatic Hydrocarbons With Elemental Sulfur and/or Pyrite Under Mild Conditions," *Energy and Fuels,* to be published.

Charles E. Schmidt[1] *and Richard F. Sprecher*[1]

Analysis of Coal-Derived Materials with Low Volatility by Low-Voltage, High-Resolution Mass Spectrometry in Conjunction with Direct-Insertion-Probe Techniques

REFERENCE: Schmidt, C. E. and Sprecher, R. F., **"Analysis of Coal-Derived Materials with Low Volatility by Low-Voltage, High-Resolution Mass Spectrometry in Conjunction with Direct-Insertion-Probe Techniques,"** *Novel Techniques in Fossil Fuel Mass Spectrometry, ASTM STP 1019,* T. R. Ashe and K. V. Wood, Eds., American Society for Testing and Materials, Philadelphia, 1989, pp. 116–132.

ABSTRACT: A low-voltage, high-resolution mass spectrometric (LVHRMS) method utilizing direct-insertion-probe techniques for analyses of coal-derived materials with low volatility is presented. The reproducibility of the LVHRMS-direct-insertion-probe method for determining heteroatomic composition, homologous series, and carbon number distributions of coal-derived materials is demonstrated. An auxiliary method to compute the experimental masses of sample peaks is described. The application of the LVHRMS-direct-insertion-probe method to the analyses of coal extracts treated with molten caustic to remove organosulfur compounds is reported. The LVHRMS-direct-insertion-probe method described can be applied to other complex materials containing heteroatoms.

KEY WORDS: low-voltage, high-resolution mass spectrometry, direct-insertion-probe techniques, coal extracts, organosulfur removal, molten caustics, mass spectrometry, fossil fuels

Mass spectrometry is often selected for the analysis of complex mixtures. Clearly, more chemical information, such as elemental analyses, carbon number distributions, heteroatom content, and number average molecular weights of high- to medium-volatile samples, can be obtained from a single low-voltage, high-resolution mass spectrometric (LVHRMS) analysis than from any other single analytical method. To date, most of the applications of LVHRMS to the analysis of fossil fuels have involved the use of heated expansion-volume inlet systems and have been limited to samples with boiling temperatures up to 650°C (1200°F) [*1–4*].

Materials having low volatility are difficult to analyze by mass spectrometry. However, the direct-insertion-probe method of admitting samples into the ion source greatly enhances the capability of detecting higher molecular weight components. Most of the applications of high-resolution mass spectrometry using direct-insertion-probe techniques have been on the analysis of pure, or simple mixtures of, materials of biological interest. A few reports on the application of direct-insertion-probe sample introduction methods to the analysis of fossil fuels by high-resolution mass spectrometry have appeared in the lit-

[1] Research chemist and research chemist, respectively, U.S. Department of Energy, DOE/PETC, P.O. Box 10940, Pittsburgh, PA 15236.

erature. Schronk et al. [*5*] utilized a microdistillation technique, based on earlier work by Grigsby and coworkers [*6*], to analyze high-boiling petroleum distillates. In this study, the sample was distilled from the direct-insertion probe into the mass spectrometer ion source by maintaining a constant ion current through a feedback circuit that ultimately controlled the probe temperature. Scheppele et al. [*7*] applied the constant ion current technique to the analysis of high-molecular-weight compounds in fossil fuels. These investigations employing microdistillation of the sample to obtain spectra of high-molecular-weight components are the only reported studies that obtained both qualitative and quantitative data on samples with the complexity of fossil fuels using direct-insertion-probe methods.

The objective of this report is to describe another approach for the analysis of high-molecular-weight components in fossil fuels. The method employs a direct-insertion probe to vaporize the sample in conjunction with low-voltage, high-resolution mass spectrometry. The approach is based on combining mass spectra taken at several different probe temperatures from 100 to 350°C. The reproducibility of the technique is demonstrated on a coal extract. This paper also reports the application of the direct-insertion-probe method to the analysis of coal extracts that had been treated with molten caustic to remove organosulfur compounds.

Procedure

Equipment—All mass spectra were obtained on a Kratos MS-50 high-resolution mass spectrometer interfaced to a Kratos DS-55 data system. The direct-insertion probe was supplied by Kratos.

Mass Spectrometer Operating Conditions—The mass spectrometer was operated at one part in 25 000 dynamic resolving power. The ionizing voltage was kept at about 11.5 eV to enhance the detection of molecular ions. The ionizing voltage was set before each determination by admitting *m*-xylene into the mass spectrometer and lowering the ionizing potential until the m/z 91 fragment was 10% of the molecular ion at m/z 106. The calibration blend and the resolution requirements for the analysis of coal products have been presented elsewhere [*4*]. The instrument was scanned from mass 700 to 60 at a rate of one mass decade/300 s. The ion source temperature was kept at 300°C.

Direct-Insertion Probe Technique—About 1 mg of sample was placed in a 3-mm capillary tube that was sealed at one end. Spectra were recorded at 100, 200, 300, and 350°C. A spectrum was also obtained before the probe was inserted into the ion source but after it was in the evacuated space. Care was taken to use the same heating pattern for all of the samples. The probe temperatures were manually set for each determination.

Coal Extract Preparation—The coal extract used in the reproducibility studies was obtained from Illinois No. 6 River King coal using a dialysis [*8*] extraction technique with a 80/20 (volume/volume) mixture of toluene and pryidine. The Bevier coal extraction procedure has been described earlier [*9*].

Molten-Caustic Treatment of Coal Extracts—The coal extract was treated with a 60/40% by-weight mixture of potassium hydroxide/sodium hydroxide (KOH/NaOH) at 375°C for 30 min. The extract was also reacted with KOH both at 375°C for 30 min and at 400°C for 3 h. In each experiment, about 100 mg of the Bevier coal extract was heated with three grams of powdered caustic in a 12.7 mm (½-in.) inside diameter Monel tubing union. The reaction products were neutralized with 3M hydrochloric acid (HCl) and then filtered. The methylene chloride soluble portion of the treated extract was analyzed by LVHRMS with direct-insertion-probe techniques. A complete description of the reaction technique has been published [*10*].

Results and Discussion

The choice of a scan rate of one mass decade/300 s warrants further discussion. The complexity of fossil fuels has been well documented [*2,4,11*]. Therefore, the highest resolving power possible should be used to analyze fossil fuels by mass spectrometry. Sensitivity is inversely related to resolving power. According to Chapman [*12*], N is proportional to $1/R^2$, where N is the number of ions in a given peak, and R is the resolving power. The peak width, t_p, can be calculated when the magnetic scan is exponential by

$$t_p = (t_{10}/2.303R) \cdot 10^6$$

where t_p is the width of a given peak in microseconds, t_{10} is the time required to scan one mass decade, and R is the resolving power. It follows that N is proportional to t_p^2, or the slower the scan rate, the greater the number of ions in a given peak. If the scan rate is too slow, the spectrum will not be representative of the sample owing to the loss of volatile material in the early part of the scan. This investigation determined that a scan rate of one mass decade/300 s was the best trade-off between making accurate mass measurements and obtaining representative mass spectra of the volatile material.

Recording mass spectra at uniform probe temperature intervals provides a pseudo-distillation of the sample into the ion source. Analysis of the resultant spectra showed that the mass-measuring accuracy, which was indicated by the root mean square (rms) [*4*] of the deviations between the measured mass and the computed mass, decreased as higher molecular weight components were being vaporized into the ion source. There are two reasons for the decrease in mass-measuring accuracy. At higher mass, greater resolving power is required to separate multiplets, and mass intervals between the calibration standards are larger at high mass. The calibration blend used in this study has six reference standards in the mass interval from 200 to 300, compared to only four reference standards for the 300 to 400 mass interval. Consequently, in the 300 to 400 mass region, where a greater number of multiplets can occur, fewer reference peaks are available for interpolating the masses of the sample peaks. To alleviate this problem, a mass calibration program, called RETROFIT, was developed to recalculate sample masses over selected mass intervals.

The program utilizes both the scan file, which contains mass estimates, and the raw data file, which contains time centroid data. The program decodes the packed scan file, selects data of interest, and sorts them into reference and sample arrays. The time centroids for selected peaks are then retrieved from the raw data file and stored in appropriate parallel arrays. If possible, two reference peaks on each side of the selected mass range are stored, along with any reference peaks in the mass range of interest. The reference data are fit to a polynomial of the following form

$$\ln(\text{reference mass}) = A + Bt + Ct^2 + Dt^3 \ldots$$

The degree of the polynomial is set to one less than the number of reference peaks used in the fit. A set of normal equations is constructed from the time and mass reference data, and solved for the coefficients by substitution using the Fortran program published by Pennington [*13*]. The masses of the sample peaks are calculated from the coefficients and the time centroids, t, according to the following relationship

$$\text{mass} = \exp(A + Bt + Ct^2 + Dt^3 \ldots)$$

and presented in tabular form. The operator may then select sample peaks to use as additional reference peaks. The selected centroids and their assigned masses are moved to the reference set, and the fitting process is repeated. When a satisfactory fit has been obtained, the scan file may be altered to reflect the new calibration.

The inclusion of four peaks outside the selected mass range assures that sufficient data will be available for a cubic fit in the region of interest. In the estimation of Bowen et al. [*14*], a cubic fit results in the best interpolation accuracy. A similar strategy was adopted by Klimowski et al. [*15*]. A minor change to the program would allow a least squares cubic fit of the reference data if more than four reference peaks were to be used, which is an alternate strategy proposed by Bowen [*14*]. We have elected to use exact fits of the reference data in order to use the highest degree polynomial possible, in accord with a suggestion by Scheppele et al. [*16*]. The program has been integrated into the Kratos DS-55 system through the use of the Kratos common-block structure and Kratos scan queue, packing and unpacking protocols.

In samples such as coal extracts and coal liquids, intense homologous series of hydrocarbons and oxygenates are present. Once these series are identified, higher homologs can be used as mass calibrants to refine and improve the overall mass-measuring capability of the LVHRMS method. The utility of RETROFIT in improving the mass determination of sample peaks is demonstrated by comparing the rms of sample peaks in the 300 to 400 mass range of the mass spectrum obtained at the 300°C probe temperature scan of the LVHRMS analysis of the Illinois No. 6 dialyzate (Table 1). For this comparison, only reference peaks at m/z 294, 308, 334, 362, 394, and 456 were included in the calibration reference file used by the DS-55 data system and RETROFIT. For the masses computed solely by the DS-55 software, the rms value for the 300 to 400 mass range was 8.93 ppm for all masses detected that had chemical significance with regard to the composition of a coal extract (isotopic peaks were omitted). When RETROFIT was applied to recompute the masses, the rms improved to 6.93 ppm. The rms for masses in the 300 to 400 mass range was further improved to 6.45 ppm by using sample peaks as intermediate standards between the calibration reference masses.

More important than the nearly 30% improvement in the mass measuring accuracy afforded by RETROFIT were the changes in the number of formula class homologs identified from the DS-55 computed masses as compared to those identified from masses recomputed by RETROFIT. As shown in Table 1, 57 homologs having the general formula, C_xH_y, were identified in the DS-55 computed masses for the 300 to 400 mass range. The identified hydrocarbon homologs increased to 83 when sample masses were used as standards. The criteria for selecting molecular formulas from precise mass determinations are listed elsewhere [*4*]. Most of the molecular ions that were identified as belonging to the C_xH_yS formula class in the DS-55 interpolation were correctly changed to C_xH_y formulas owing to the more accurate mass determinations afforded by using RETROFIT with sample peaks as additional references. The number of homologs in the monooxygenate formula class (C_xH_yO) also increased as a result of using RETROFIT to calculate the masses of the sample peaks. The other formula class that changed significantly was the $C_xH_yNO_2$ homologous series. When RETROFIT was used to calculate the sample peak masses, 16 of the species identified from the DS-55 interpolation as $C_xH_yNO_2$ were correctly identified as carbon-13 isotope molecular ions from hydrocarbons. In fact, the intensities of the molecular ions identified as carbon-13 isotopes by RETROFIT corresponded to the natural abundance of the respective hydrocarbons. In general, RETROFIT improved the accuracy of the mass determinations, thereby removing some of the uncertainties in assigning chemical formulas to the measured masses and resulting in more meaningful formula selection. Also given in Table 1 are the elemental analyses computed for the three mass determination procedures. The elemental analyses were calculated assuming unit sensitivities for all the molecular ions detected using computer programs previously described [*4*]. The chemical composition of the molecular ions in the 300 to 400 mass range determined by

TABLE 1—*Comparison of DS-55 and RETROFIT interpolation methods for the mass range of 300 to 400 for the 300°C scan for the Illinois No. 6 coal dialyzate.*

	Number of Homologs Detected		
Formula Class	DS-55	RETROFIT	RETROFIT + Sample Peaks
C_xH_y	57	80	83
C_xH_yO	49	60	62
$C_xH_yO_2$	33	38	40
$C_xH_yO_3$	14	14	14
C_xH_yS	19	7	3
C_xH_yOS	8	4	2
$C_xH_yO_2S$	5	2	0
C_xH_yN	13	14	15
C_xH_yNO	5	5	5
$C_xH_yNO_2$	18	5	3
$C_xH_yNO_3$	5	0	0
rms[a]	8.93	6.93	6.45
C[b]	85.7	88.3	88.5
H	8.44	8.12	8.10
N	0.569	0.182	0.172
O	3.92	3.13	3.16
S	1.33	0.266	0.110

[a] The rms data are given in ppm.
[b] The elemental analyses are given in percent.

LVHRMS methods for the 300°C probe spectrum changed significantly depending on the mass interpolation scheme employed. As can be seen in Table 1, elemental sulfur and nitrogen values were reduced considerably when RETROFIT was used in conjunction with designated sample peaks as standard references because C_xH_yS molecular ion identifications were changed to C_xH_y formulas, and $C_xH_yNO_2$ species were correctly identified as isotopic carbon-13 molecular ions of hydrocarbons.

With regard to acquiring qualitative information on intractable samples, such as coal liquid residua, high-boiling petroleum distillates, and other complex materials with boiling temperatures above 343°C (650°F), the direct-insertion-probe method of sample introduction expands the analytical capabilities of mass spectrometry. However, quantitative data on the chemical composition of samples admitted into the ion source by the direct-insertion probe are difficult to obtain because of the instability of the ion current. Also, the resulting spectra may not be representative of the sample owing to loss of lower molecular weight components. Schronk et al. [5] addressed the problem of ion current fluctuations by maintaining constant ion current through adjustment of the probe temperature during analysis. Unless one records the entire spectrum instantaneously, such as in focal plane instruments employing photoplates for ion detection, the problem of not analyzing lower molecular weight components when the probe temperature is increased still exists. The problem is especially severe in mass spectrometers that scan from high mass to low mass, as the MS-50 does.

Semiquantitative data on the chemical composition of high-boiling materials can be obtained by heating the sample at selected probe temperatures, recording the mass spectrum at each different probe temperature, and combining the spectra. In this fashion, the sample is distilled into the ion source, and spectra are recorded at constant probe temperatures for each selected probe temperature. Each spectrum then is treated as a single analysis, and the molecular ion intensities are combined for all six scans (see the Procedure section). Data reduction techniques are then applied to the composite spectrum, yielding carbon number distributions of each structural series detected and semiquantitative data based on the sum of the molecular ion intensities. Since the spectra are recorded at low ionizing voltage, one can calculate relative mole percentages of each molecular formula detected, as well as number average molecular weights and average Z numbers (a measure of hydrogen deficiency) for each formula class and for the total sample.

The aforementioned data are obtained using unit sensitivities for all molecular species detected. Computing the relative mole percent concentration of molecular ions in this fashion ignores the effect of the size of the aromatic nucleus on the cross section for ionization and the variation in ionization potential of the compounds of interest. Normally, molecular ion intensities are corrected for differences under LVHRMS conditions by applying predetermined relative mole or gram sensitivity correction factors. Because no corrections are made for the large differences in low-voltage sensitivity between single-ring and multiring aromatic structures, the relative mole percent data are biased toward the higher molecular weight components. However, relative mole percent data are useful when comparing similar samples.

Shown in Table 2 are the reproducibilities of formula class determinations and related parameters of a pyridine dialyzate from Illinois No. 6 River King coal. The data represent the average of three direct-insertion-probe analyses of the dialyzate. About 45% of the sample was vaporized in each determination. The reproducibility data for formula class determinations can be divided into two general categories: (1) those with standard deviations less than 10% of the average relative mole percent, and (2) those with standard deviations greater than 10%. Apparently, the reproducibility data are more dependent on the resolution requirements of the sample than on the analytical method used to obtain the mass spectra. For the C_xH_y, C_xH_yO, and $C_xH_yO_2$ formula classes, the resolution needed to separate multiplets commonly found in coal products is not as high as that required to separate mass multiplets that contain sulfur and nitrogen. Thus, the standard deviations for the C_xH_y, C_xH_yO, and $C_xH_yO_2$ formula classes are less than 10% of the average value. The relative abundances of each respective class also have a bearing on the reproducibility. The standard deviations for the C_xH_yN and C_xH_yNO formula classes were slightly less than 10%

TABLE 2—*Average relative mole percent ($\overline{X}$) and standard deviations (σ) for formula classes and related parameters of a pyridine dialyzate from Illinois No. 6 River King coal determined by LVHRMS utilizing direct-insertion-probe techniques.*

Formula Class	$\overline{X}$	σ
Hydrocarbons	33.28	2.82
Monooxygen compounds	27.66	1.37
Dioxygen compounds	16.92	1.66
Trioxygen compounds	4.68	0.56
Nitrogen compounds	5.27	0.46
Nitrogen-oxygen compounds	4.18	0.37
Nitrogen-dioxygen compounds	3.03	0.86
Sulfur compounds	2.44	0.91
Sulfur-oxygen compounds	1.44	0.08
Sulfur-dioxygen compounds	0.36	0.22
Molecular weight range	95 to 486	
Number average molecular weight	283	7.75

NOTE—Values of $\overline{X}$ and σ are based on three determinations.

of the average relative mole percent for each of the two classes. At a resolution of one part in 25 000, the maximum mass at which the ^{13}CH-N doublet can be base-line resolved is 203. Therefore, for the pyridine dialyzate, with a number average molecular weight of 283 (as shown in Table 2), many of the nitrogen-containing multiplets would be resolved. The $C_xH_yNO_2$ and the sulfur-containing formula classes all had standard deviations greater than 10% of their average relative mole percents except for the C_xH_yOS formula class. The excellent standard deviation of 0.08% is no doubt a result of the small number of C_xH_yOS compounds detected. Overall, the reproducibility of the direct-insertion-probe technique coupled with low-voltage, high-resolution methods shown in Table 2 indicates the technique is acceptable for the analysis of similar samples.

The analytical power of the LVHRMS direct-insertion-probe technique is illustrated in Table 3. The compound types determined in the pyridine dialyzate from Illinois No. 6 coal are summarized according to the method of Aczel et al. [*17*]. The distribution of the components in the dialyzate is listed according to each formula class detected. Also listed for each formula class in Table 3 are the number and range in Z of homologous series with the general formula C_nH_{2n+Z}. The total number of homologs and their carbon-number range are also reported for each formula class.

The most abundant formula class was the hydrocarbon (C_xH_y) class with 27 homologous (Z) series and 275 individual homologs. Since Z is related to the number of aromatic carbons (C_a) and rings (R) by the relationship $Z = 2 - (C_a + 2R)$, each Z series corresponds to a different structural type. Thus, for the hydrocarbons, 27 distinct structural series were identified in the coal dialyzate. The Z range of 0 to -52 would include structures having one saturated ring to those containing up to eight aromatic rings. Each molecular formula (homolog) within a specific Z series includes structural isomers, such as pyrene and fluoranthene in the $Z = -22$ series, as well as positional isomers for the alkyl substituted homologs. Thus, there are many more compounds in the hydrocarbon class than indicated

TABLE 3—*Summary of compound types determined in a pyridine dialyzate from Illinois No. 6 coal.*

		Homologous Series		Homologs	
Formula Class	Relative Mole Percent	Number	Z Number Range	Number	Carbon Number Range
Hydrocarbons	33.28	27	0 to 52	275	6 to 37
Monooxygenates	27.66	22	6 to 48	195	7 to 35
Dioxygenates	16.92	19	8 to 46	146	8 to 34
Nitrogenates	5.27	20	11 to 43	100	11 to 33
Trioxygenates	4.68	15	12 to 40	94	13 to 33
Nitrogen-oxygen compounds	4.18	19	11 to 45	92	10 to 33
Nitrogen-dioxygen compounds	3.03	17	7 to 43	71	11 to 31
Sulfur compounds	2.44	16	10 to 44	46	9 to 31
Sulfur-oxygen compounds	1.44	14	12 to 40	50	11 to 31
Miscellaneous N, O, S compounds	0.74	12	7 to 33	36	16 to 31
Sulfur-dioxygen compounds	0.36	9	12 to 34	23	16 to 31
Totals	100.00	190	0 to 52	1128	6 to 37

TABLE 4—*Summary of formula class distributions of treated and untreated Bevier coal extracts.*

Formula Class	Untreated Coal Extract	Treated Coal Extracts: 375°C, 30 min KOH/NaOH (60/40)	Treated Coal Extracts: 375°C, 30 min KOH	Treated Coal Extracts: 400°C, 3 h KOH
C_xH_y	32	16	48	57
C_xH_yO	21	44	28	23
$C_xH_yO_2$	7.6	14	4.5	3.3
$C_xH_yO_3$	2.3	0.92	0.57	0.16
C_xH_yS	16	5.5	3.5	2.7
C_xH_yOS	9.4	6.0	2.5	1.2
$C_xH_yO_2S$	3.6	3.8	0.14	0.95
C_xH_yN	3.9	3.7	8.3	9.5
C_xH_yNO	2.8	4.9	3.1	1.6
C_xH_y, C_xH_yNOS, $C_xH_yS_2$, C_xH_yNS	1.1	0.83	0.70	. . .

by the 275 homologs identified by LVHRMS. The carbon number range for the hydrocarbon formula class was C_6–C_{37}. Oxygenated compounds (C_xH_yO, $C_xH_yO_2$, and $C_xH_yO_3$) accounted for nearly 50% of the components in the dialyzate. In summary, molecular formulas corresponding to 190 homologous series with a Z range of 0 to -52 were found in the Illinois No. 6 pyridine dialyzate. The overall range in carbon number was C_6–C_{37}. Analysis of the pyridine dialyzate from Illinois No. 6 River King coal by direct-insertion-probe LVHRMS methods yielded relative mole percent determinations for 1128 individual homologs.

The direct-insertion-probe method of analyzing samples with low volatility by mass spectrometry just described was applied to samples of coal extracts that had been treated with molten caustic to remove sulfur. In this study, pyridine extracts of a Bevier seam coal were subjected to treatment with molten alkali. Different caustic mixtures and reaction times and temperatures were used. Our goal in analyzing the molten-caustic-treated coal extracts by LVHRMS direct-insertion-probe techniques was to determine the organosulfur removal by molten caustic. A summary of the formula class distributions for both molten-caustic-treated and the untreated Bevier coal extracts is given in Table 4. Thirteen formula classes were determined by LVHRMS direct-insertion-probe techniques for the caustic-treated Bevier coal extracts and the original Bevier coal extract. Relative mole percent determinations for each homologous Z series in each formula class except the $C_xH_yNO_2$, C_xH_yNS, and $C_xH_yS_2$ classes, and the carbon number range within each Z series, are given in Tables 5 through 13 in the Appendix for the treated and untreated Bevier coal extracts.

The sulfur-containing components that have the general formula C_xH_yS decreased from 16% for the raw coal extract to 5.5% for the extract treated with a 60/40% by weight mixture of molten KOH/NaOH at 375°C for 30 min. The C_xH_yS species were further reduced to a value of 3.5% when the coal extract was treated with only KOH at 375°C for 30 min. When the coal extract was treated with KOH at 400°C for 3 h, the C_xH_yS components decreased from the initial amount in the coal extract (16%), to 2.7%—an 83% reduction. Among the most abundant homologous series in the C_xH_yS formula class were thiophenes ($Z = -4$), benzothiophenes ($Z = -10$), and dibenzothiophenes ($Z = -16$), as shown in Table 9 in the Appendix. These results are in agreement with earlier published accounts of the organosulfur compounds in the Bevier coal extract [*9*], as determined by gas chromatography/mass spectrometry and gas chromatography with flame photometric detection specific for sulfur.

Another sulfur-containing formula class, C_xH_yOS, also decreased from 9.4% in the raw coal extract to 6.0% for the extract treated with a 60/40% by weight mixture of KOH/NaOH for 30 min at 375°C. The levels of C_xH_yOS components were further reduced to 1.2% for the extract treated with only KOH at 400°C for 3 h. Among those compounds showing large decreases, as a result of molten-caustic treatment, were structures with molecular formulas corresponding to hydroxynaphthenobenzothiophenes ($Z = -12$), hydroxynaphthenothiophenes ($Z = -6$), and hydroxynaphthenodibenzothiophenes ($Z = -18$) (see Table 10 in the Appendix).

The fraction of hydrocarbons (C_xH_y) decreased from 32% in the untreated pyridine extract to 16% for the extract treated with the 60/40% by weight mixture of molten hydroxide. A large increase in the level of hydrocarbons (57%) compared to the raw coal extract (32%) was observed for the extract treated solely with KOH at 400°C for 3 h. In a previous investigation [*9*] of the organosulfur substituents in Bevier seam coal, an organosulfur-rich concentrate, isolated from the Bevier coal, contained predominately heterocyclic sulfur compounds having mainly benzothiophene and dibenzothiophene structures. Model compound studies [*10*] have shown the desulfurization of benzothiophene by molten hydroxide occurs in two steps. First, initial cleavage of the thiophene ring results in the loss of a 1-carbon fragment, leaving the salt of an aromatic thiol. Subsequently, sulfur is lost from the aromatic thiol, resulting in the formation of toluene. The increase in the fraction of hydrocarbons for the 375°C treatment with KOH over the amount in the untreated extract, and the increase in the 400°C-treated extract compared to that at 375°C with KOH, are consistent with the benzothiophene model compound study. However, the decrease in hydrocarbon content of the extract treated at 375°C with KOH/NaOH is not consistent with the model compound study or the KOH-only treatment and remains unexplained at this time.

The amounts of monooxygenated compounds (C_xH_yO) also varied considerably upon treatment of the Bevier coal pyridine extract with caustic under different conditions. The reaction of the extract with the 60/40 melt at 375°C for 30 min resulted in a doubling of the C_xH_yO compounds, from 21% of the extract to 44% of the treated extract. The concentration of oxygenates (C_xH_yO) remained about the same for the reactions with 100% KOH at 375°C for 30 min and 400°C for 3 h, when compared to the untreated Bevier coal extracts. Tables 6 and 7 in the Appendix contain the various homologous series of C_xH_yO and $C_xH_yO_2$, respectively, for the treated and untreated Bevier extracts. The most striking feature in Table 6 is the large increase in the phenol series ($Z = -6$), from 0.67% of the untreated coal extract to 17% of the 60/40 KOH/NaOH molten-caustic-treated extract. Similarly, the naphthols ($Z = -12$) and phenanthrols or anthrols or both ($Z = -18$) increased in the extract treated with the KOH/NaOH mixture. The levels of the aforementioned mono-oxygenates were slightly higher in the Bevier coal extracts treated with KOH only than in the untreated Bevier coal extract. For the dioxygenates (Table 7), the $Z = -14$, -16, and -18 homologs, corresponding to dihydroxyacenaphthenes, dihydroxyacenaphthylenes, and dihydroxyphenananthrenes, respectively, increased the most upon treatment of the coal extract with the 60/40% by weight mixture of molten KOH/NaOH. Little change was noticed in the KOH-only treated coal extracts as compared to the raw-coal extract.

The fraction of nitrogen-containing compounds, formula class C_xH_yN, increased when the coal extract was treated with molten KOH. No explanation for such an increase can be given at this time. One possible reason could be that the increase is an artifact caused by basing the relative mole percent values on total ionization. However, an increase in elemental nitrogen in coals treated with molten caustic has been reported [*18*], but no explanation was given for the increase.

Conclusions

A low-voltage, high-resolution mass-spectrometric technique, in conjunction with direct-insertion-probe sample vaporization methods, has been developed for analysis of samples with low volatility. Quantitative information expressed as relative mole percent concentrations can be obtained for both hydrocarbon and heteroatomic constituents of coal-derived materials, yielding the carbon number distributions, the *Z* number ranges, and the distributions of distinct homologous series. The chemical significance of the molecular formulas determined by the LVHRMS direct-insertion-probe method was improved by using an auxiliary method to compute the experimental masses of the sample peaks. The overall effectiveness of the LVHRMS direct-insertion probe method described herein could be enhanced by correcting the intensities of the components detected for sensitivity differences at LVHRMS direct-insertion probe conditions.

Thiophenic sulfur moieties were selectively removed from a coal extract by molten caustic, as determined by LVHRMS direct-insertion-probe methods. Even though the concentrations of the C_xH_yOS and $C_xH_yO_2S$ formula classes were less than the standard deviations of the technique, in general, the levels of all sulfur-containing oxygenates decreased as a result of molten-caustic treatment, while those of the nonsulfur-containing oxygenates increased. The increase in oxygenates in the molten-caustic treatment of the Bevier coal extracts appears to be related to the concentration of NaOH in the caustic melt.

The large decrease in the fraction of sulfur compounds (C_xH_yS), from 16% in the Bevier coal extract to 5.5% for the extract treated with the 60/40 blend of KOH/NaOH, seems to indicate that a large portion of organosulfur compounds is readily removed under those conditions. More severe reaction conditions (100% KOH at 400°C for 3 h) were required to further reduce the level of sulfur compounds in the Bevier coal extract.

Acknowledgments

The authors are indebted to Joseph Malli, Jr., for obtaining the mass spectra and to Dr. Bruce R. Utz for treating the coal extracts with molten caustic. The authors also are indebted to Dr. Dennis H. Finseth for the pyridine dialyzate of the Illinois No. 6 River King coal.

Reference in this report to any specific commercial product, process, or service is to facilitate understanding and does not necessarily imply its endorsement or favoring by the U.S. Department of Energy.

APPENDIX

Relative Mole Percent Determinations for the Raw and Molten-Caustic-Treated Bevier Coal Extracts

TABLE 5—*LVHRMS data for C_xH_y formula class.*

			Treated Coal Extracts					
			375°C, 30 min				400°C, 3 h	
	Untreated Coal Extract		KOH/NaOH (60/40)		KOH		KOH	
−Z	Carbon Number Range	Relative Mole Percent	Carbon Number Range	Relative Mole Percent	Carbon Number Range	Relative Mole Percent	Carbon Number Range	Relative Mole Percent
0			6	0.16	5 to 12	0.60	7 to 9	0.10
2			6	<0.10	6 to 29	1.6	6 to 10	0.65
4	8 to 9	0.18	10	<0.10	7 to 29	0.85	7 to 15	0.41
6	6 to 8	0.37	6 to 10	0.54	6 to 31	2.3	6 to 22	3.3
8	12 to 14	0.41	9	<0.10	9 to 30	1.4	9 to 21	2.5
10	11 to 18	0.62	12 to 14	<0.10	10 to 30	0.97	9 to 19	1.0
12	10 to 19	5.4	11 to 16	3.7	10 to 30	8.5	10 to 22	12
14	13 to 20	3.5	12 to 18	2.0	12 to 30	6.0	12 to 27	8.2
16	12 to 23	2.9	12 to 19	1.6	12 to 30	3.5	13 to 29	5.2
18	14 to 28	7.1	14 to 22	2.6	14 to 30	6.1	14 to 23	7.6
20	15 to 21	2.2	16 to 21	0.69	15 to 30	3.1	16 to 24	2.9
22	16 to 26	3.0	16 to 23	2.2	16 to 29	5.0	16 to 26	5.2
24	18 to 23	1.9	18 to 23	0.85	18 to 28	3.1	18 to 26	2.6
26	18 to 29	1.2	21 to 25	0.13	19 to 28	1.1	20 to 26	0.85
28	20 to 26	1.6	21 to 26	0.90	20 to 28	1.9	20 to 25	3.0
30	22 to 26	0.49	22 to 26	0.22	22 to 27	0.88	23 to 26	0.81
32	23 to 26	0.51	22 to 27	0.27	23 to 30	0.76	23 to 27	0.46
34	24 to 37	0.36	24 to 26	0.17	25 to 28	0.37	26 to 27	0.11
36	26 to 27	0.14	27 to 28	0.11	30	<0.10	28	0.03
38	26 to 29	0.24			28 to 29	<0.10		
40	28	<0.10						
42	...	...						
44	31	<0.10						

TABLE 6—*LVHRMS data for C_xH_yO formula class.*

	Untreated Coal Extract		Treated Coal Extracts					
			375°C, 30 min				400°C, 3 h	
			KOH/NaOH (60/40)		KOH		KOH	
−*Z*	Carbon Number Range	Relative Mole Percent	Carbon Number Range	Relative Mole Percent	Carbon Number Range	Relative Mole Percent	Carbon Number Range	Relative Mole Percent
6	7 to 8	0.67	6 to 11	17	6 to 12	2.9	7 to 11	3.0
8	...	...	9 to 14	3.6	8 to 16	1.3	9 to 14	0.36
10	11	<0.10	9 to 13	1.3	9 to 16	0.65	10 to 14	0.19
12	...	...	10 to 18	4.3	10 to 18	3.1	11 to 13	0.62
14	15	0.39	12 to 17	2.7	12 to 19	2.9	12 to 19	8.9
16	13 to 21	2.6	13 to 19	2.6	13 to 22	3.7	13 to 18	1.6
18	15 to 20	1.7	14 to 20	4.0	14 to 20	3.6	14 to 21	1.7
20	16 to 22	1.0	16 to 21	2.2	16 to 23	2.8	16 to 21	1.9
22	16 to 24	4.3	16 to 23	2.2	16 to 25	2.7	17 to 24	1.3
24	18 to 24	2.4	19 to 23	1.4	18 to 26	1.7	17 to 23	1.7
26	19 to 26	0.98	20 to 25	0.58	19 to 28	0.84	20 to 23	0.71
28	20 to 25	2.6	20 to 24	1.1	20 to 27	0.86	20 to 24	0.38
30	22 to 29	1.1	23 to 27	0.66	22 to 28	0.57	21 to 24	0.42
32	22 to 29	0.8	24 to 27	<0.10	23 to 29	0.39		
34	23 to 30	1.1	24 to 27	0.22	25 to 29	0.19		
36	25 to 31	0.61	26 to 29	0.12	27 to 30	<0.10		
38	27 to 32	0.47			28	<0.10		
40	28 to 31	0.18						
42	29 to 30	0.16						
44	30	<0.10						

TABLE 7—*LVHRMS data for $C_xH_yO_2$ formula class.*

	Untreated Coal Extract		Treated Coal Extracts					
			375°C, 30 min				400°C, 3 h	
			KOH/NaOH (60/40)		KOH		KOH	
−Z	Carbon Number Range	Relative Mole Percent	Carbon Number Range	Relative Mole Percent	Carbon Number Range	Relative Mole Percent	Carbon Number Range	Relative Mole Percent
6	7	<0.10	6 to 8	0.42	6 to 7	<0.10		
10	...	...	10	<0.10	10	<0.10		
12	...	...	11 to 12	0.18	11 to 12	0.10		
14	...	...	12 to 18	3.8	13 to 18	1.1	12 to 17	2.6
16	14 to 15	0.21	13 to 19	2.3	13 to 20	0.57	14	<0.10
18	14 to 17	0.32	14 to 21	2.2	14 to 19	0.63	15 to 16	<0.10
20	17 to 20	0.18	17 to 22	1.0	17 to 29	0.32	17	<0.10
22	16 to 22	2.0	17 to 24	0.90	17 to 22	0.65	18 to 22	0.45
24	19 to 24	1.0	18 to 23	1.1	19 to 24	0.42	18 to 19	<0.10
26	20 to 24	0.56	20 to 24	0.57	20 to 25	0.23	20	<0.10
28	21 to 26	0.78	21 to 25	0.47	21 to 25	0.18		
30	22 to 28	0.54	23 to 27	0.54	23 to 28	0.17		
32	23 to 28	0.91	28 to 27	0.15				
34	25 to 28	0.54						
36	25 to 28	0.16						
38	26 to 31	0.29						
40	27	<0.10						
42	31	<0.10						

TABLE 8—*LVHRMS data for $C_xH_yO_3$ formula class.*

	Untreated Coal Extract		Treated Coal Extracts					
			375°C, 30 min				400°C, 3 h	
			KOH/NaOH (60/40)		KOH		KOH	
−Z	Carbon Number Range	Relative Mole Percent	Carbon Number Range	Relative Mole Percent	Carbon Number Range	Relative Mole Percent	Carbon Number Range	Relative Mole Percent
6			6 to 8	0.42	6	0.11	6	<0.10
10			10	<0.10	...	...	...	...
12			11 to 12	0.18	14 to 17	0.27	15 to 16	0.10
14			12 to 18	3.8	15 to 16	<0.10	16	0.10
16			13 to 19	2.3	...	...		
18	19	<0.10	14 to 21	2.2	17 to 21	<0.10		
20	18	<0.10	17 to 22	1.0				
22	17 to 18	<0.10	17 to 24	0.90				
24	18 to 24	0.27	18 to 23	1.1				
26	20 to 23	0.25	20 to 24	0.57				
28	21 to 26	0.46	21 to 25	0.47				
30	24 to 29	0.25	23 to 27	0.54				
32	25 to 26	0.28	26 to 27	0.15				
34	25 to 30	0.35						
36	25 to 29	0.15						

TABLE 9—*LVHRMS data for C_xH_yS formula class.*

			Treated Coal Extracts					
			375°C, 30 min				400°C, 3 h	
	Untreated Coal Extract		KOH/NaOH (60/40)		KOH		KOH	
−Z	Carbon Number Range	Relative Mole Percent	Carbon Number Range	Relative Mole Percent	Carbon Number Range	Relative Mole Percent	Carbon Number Range	Relative Mole Percent
0	11	<0.10					13 to 15	0.19
2	10 to 13	2.3	9 to 11	0.73	14	<0.10	10	0.11
4	11 to 16	2.1	11 to 15	0.41	11 to 16	0.10	11 to 16	0.24
6	13 to 18	0.33	6 to 16	0.74	14 to 15	0.49	10 to 18	0.35
8	12 to 20	1.2	11 to 19	0.89	12 to 16	0.66	12 to 17	0.71
10	11 to 17	1.4	11 to 18	0.41	11 to 19	0.36	13 to 16	0.48
12	13 to 21	1.4	13 to 20	0.68	13 to 19	0.58	18 to 19	0.24
14	13 to 20	1.3	13 to 20	0.60	12 to 23	0.43	19	0.21
16	13 to 20	2.2	15 to 22	0.25	12 to 21	0.24	12 to 18	<0.10
18	16 to 22	1.1	18 to 23	0.49	17 to 23	0.36		
20	19 to 23	0.24	17	<0.10	20 to 21	<0.10		
22	16 to 22	0.98	19 to 23	0.21	22 to 24	0.13		
24	20 to 24	0.36	22	<0.10	24	<0.10		
26	21 to 23	0.33	25	<0.10				
28	21 to 26	0.24						
30	22 to 25	<0.10						

TABLE 10—*LVHRMS data for C_xH_yOS formula class.*

			Treated Coal Extracts					
			375°C, 30 min				400°C, 3 h	
	Untreated Coal Extract		KOH/NaOH (60/40)		KOH		KOH	
−Z	Carbon Number Range	Relative Mole Percent	Carbon Number Range	Relative Mole Percent	Carbon Number Range	Relative Mole Percent	Carbon Number Range	Relative Mole Percent
2			8 to 15	1.0	10 to 12	0.24	8	<0.10
4			9 to 13	0.84	10	0.19	11 to 14	0.11
6	11 to 18	1.1	6 to 16	0.69	13 to 14	0.16	11 to 12	<0.10
8	12 to 17	0.80	11 to 17	0.31	11 to 15	0.48	11 to 13	0.22
10	9 to 19	0.77	9 to 18	0.75	13 to 17	0.42	14 to 18	0.12
12	13 to 21	2.4	13 to 18	0.92	15 to 18	0.58	15 to 20	0.21
14	15 to 21	0.55	16 to 20	0.44	15 to 19	0.11	17 to 19	0.34
16	16 to 23	0.35	17 to 21	0.27	18 to 20	<0.10	20	<0.10
18	18 to 22	0.85	17 to 22	0.29	20 to 22	0.14		
20	15 to 26	0.70	23	<0.10	20 to 23	0.17		
22	18 to 26	0.56	21 to 22	<0.10				
24	20 to 23	0.27	23	<0.10				
26	20 to 28	0.57						
28	22 to 29	0.17						
30	23 to 28	0.24						
32	24	0.10						
34	27	<0.10						

TABLE 11—*LVHRMS data for $C_xH_yO_2S$ formula class.*

	Treated Coal Extracts							
			375°C, 30 min				400°C, 3 h	
	Untreated Coal Extract		KOH/NaOH (60/40)		KOH		KOH	
−Z	Carbon Number Range	Relative Mole Percent	Carbon Number Range	Relative Mole Percent	Carbon Number Range	Relative Mole Percent	Carbon Number Range	Relative Mole Percent
4			9 to 15	1.1	13	<0.10	11 to 13	0.62
6	11 to 12	0.10	10 to 16	0.68	...	...	...	...
8	13 to 14	0.12	11 to 18	0.75	12 to 15	<0.10	11	<0.10
10	16 to 17	<0.10	15 to 19	0.21	...	...	...	...
12	14 to 19	0.60	14 to 21	0.43	...	...	17 to 18	0.12
14	15 to 21	0.39	15 to 19	0.25	16 to 18	<0.10	...	...
16	17 to 21	0.34	18 to 20	0.21			12	0.18
18	18 to 22	0.57	18 to 19	0.10				
20	19 to 25	0.40	20	<0.10				
22	20 to 25	0.25						
24	22 to 24	0.26						
26	22 to 25	0.16						
28	22 to 28	0.24						
30	24	<0.10						
32	28	<0.10						

TABLE 12—*LVHRMS data for C_xH_yN formula class.*

	Treated Coal Extracts							
			375°C, 30 min				400°C, 3 h	
	Untreated Coal Extract		KOH/NaOH (60/40)		KOH		KOH	
−Z	Carbon Number Range	Relative Mole Percent	Carbon Number Range	Relative Mole Percent	Carbon Number Range	Relative Mole Percent	Carbon Number Range	Relative Mole Percent
5			8	0.13	6 to 9	0.27	6 to 10	0.96
7			...	...	11	<0.10	11	<0.10
9			10 to 11	0.11	10 to 11	0.11	...	...
11			12	0.10	11 to 13	0.31	10 to 12	0.20
13	12	0.32	...	...	15	<0.10	...	...
15	15 to 16	0.14	14 to 16	0.40	13 to 18	1.2	12 to 16	3.04
17	17	<0.10	14 to 18	0.38	15 to 19	0.60	15 to 17	0.18
19	...	...	...	...	16 to 21	0.43	...	...
21	17 to 21	1.5	17 to 20	0.95	16 to 22	2.2	16 to 21	3.5
23	19 to 20	0.19	20 to 22	0.40	18 to 23	0.95	18 to 21	0.47
25	24	<0.10	20	<0.10	19 to 24	0.33	20 to 22	0.11
27	20 to 25	0.81	21 to 24	0.56	20 to 25	0.98	20 to 23	0.92
29	22 to 26	0.44	24 to 25	0.20	23 to 25	0.41	23	<0.10
31	25	0.10	23 to 27	0.29	24 to 27	0.24	24	<0.10
33	26	<0.10	25 to 28	0.19	26 to 28	0.14		
35	...	...			28 to 29	<0.10		
37	26	0.10						
39	...	...						
41	28	<0.10						

TABLE 13—*LVHRMS data for C_xH_yNO formula class.*

			Treated Coal Extracts					
			375°C, 30 min				400°C, 3 h	
	Untreated Coal Extract		KOH/NaOH (60/40)		KOH		KOH	
$-Z$	Carbon Number Range	Relative Mole Percent	Carbon Number Range	Relative Mole Percent	Carbon Number Range	Relative Mole Percent	Carbon Number Range	Relative Mole Percent
5					7 to 8	<0.10		
9					10	<0.10		
11	10 to 11	0.18	10 to 13	0.81	10 to 13	0.78	11	<0.10
13	...	...	13 to 15	0.20	14 to 15	<0.10	...	...
15	...	...	14 to 18	0.15	13 to 15	<0.10	...	...
17	14 to 17	0.39	14 to 16	0.92	13 to 17	0.59	14 to 18	0.55
19	...	...	16 to 18	0.40	17 to 19	0.15	...	...
21	16 to 21	0.43	16 to 24	1.0	16 to 20	0.53	16 to 18	0.50
23	19 to 21	0.40	18 to 20	0.47	18 to 24	0.43	18 to 20	0.40
25	21 to 22	0.14	21 to 23	0.13	20 to 24	<0.10	...	...
27	20 to 23	0.52	21 to 23	0.24	20 to 23	0.23	20 to 21	<0.10
29	23 to 25	0.13	25	<0.10	23 to 24	<0.10		
31	25 to 26	0.23	22 to 27	0.29				
33	25	0.24	25 to 26	0.14				
35	...	...						
37	26	<0.10						
39	27 to 29	0.10						

References

[*1*] Aczel, T. A., *Reviews of Analytical Chemistry,* Vol. 1, No. 3, March 1971, pp. 226–261.

[*2*] Aczel, T., Williams, R. B., Pancirov, R. J., and Karchmer, J. H., "Chemical Properties of SYNTHOIL Products and Feeds," MERC-8007-1, Parts 1 to 5, Sept. 1976.

[*3*] Schmidt, C. E., Perry, M. B., White, C. M., Gibbon, G. A., and Retcofsky, H. L., "Structural Definition of Synthetic Fuels: H-Coal Distillates," DOE/PETC/TR-87/4, U.S. Department of Energy, 1986.

[*4*] Schmidt, C. E., Sprecher, R. F., and Batts, B. D., *Analytical Chemistry,* Vol. 59, No. 17, Sept. 1987, pp. 2027–2033.

[*5*] Schronk, L. R., Grigsby, R. D., and Scheppele, S. E., *Analytical Chemistry,* Vol. 54, No. 4, April 1982, pp. 748–755.

[*6*] Grigsby, R. D., Hansen, C. O., Mannering, D. G., Fox, W. G., and Cole, R. H., *Analytical Chemistry,* Vol. 43, No. 8, July 1971, pp. 1135–1137.

[*7*] Scheppele, S. E., Grindstaff, Q. G., Grigsby, R. D., Chung, K. C., Hwang, C. S., Marriott, T. D., and Schronk, L. R., "Automated Probe Microdistillation/Mass Spectrometry for the Analysis of High-Molecular-Weight Compounds in Fossil Fuels," DOE/BETC/TPR-82/1, U.S. Department of Energy, 1982.

[*8*] Finseth, D. H., Prozcuek, B. J., and Koppenaal, D. W., *Fuel,* Vol. 61, No. 11, 1982, pp. 1155–1159.

[*9*] White, C. M., Douglas, L. J., Perry, M. B., and Schmidt, C. E., *Journal of Energy and Fuels,* Vol. 1, No. 2, March/April 1987, pp. 222–226.

[*10*] Utz, B. R., Soboczenski, S. K., and Friedman, S., *American Chemical Society Division of Fuel Preprints,* Vol. 30, No. 2, April/May 1985, pp. 35–40.

[*11*] Scheppele, S. E., Benson, P. A., Greenwood, G. J., Grindstaff, Q. G., Aczel, T. A., and Beier, B. F., *Advances in Chemistry Series,* Vol. 195, 1981, pp. 53–82.

[*12*] Chapman, J. R., *Computers in Mass Spectrometry,* Academic Press, London, 1978, p. 63.
[*13*] Pennington, R. H., *Introductory Computer Methods and Universal Analysis,* The Macmillan Co., New York, 1965, pp. 342–345.
[*14*] Bowen, H. C., Clayton, E., Shields, D. J., and Stanier, H. M., *Advances in Mass Spectrometry,* Vol. 4, E. Kendrick, Ed., Institute of Petroleum, London, 1968, p. 257.
[*15*] Klimowski, R. J., Venkataraghavan, R., and McLafferty, F. W., *Journal of Organic Mass Spectrometry,* Vol. 4, 1970, pp. 17–39.
[*16*] Scheppele, S. E., Grindstaff, Q. G., Grigsby, R. D., McDonald, S. R., and Hwang, C. S., *International Journal of Mass Spectrometry and Ion Physics,* Vol. 49, 1983, pp. 179–209.
[*17*] Aczel, T., Williams, R. B., Chamberlain, N. F., and Lumpkin, H. E., *Advances in Chemistry Series,* Vol. 195, 1981, pp. 237–251.
[*18*] "Bench-Scale Development of the Gravimelt Process for Cleaning Coal," Final Report, Contract No. DE-AC22-83PC63032, U.S. Department of Energy, Aug. 1987.

Richard W. Crawford,[1] Thomas T. Coburn,[1] Philip E. Miller,[1] and Myongsook S. Oh[1]

On-Line, Mass Spectrometric Determination of Ammonia from Oil Shale Pyrolysis Using Isobutane Chemical Ionization

REFERENCE: Crawford, R. W., Coburn, T. T., Miller, P. E., and Oh, M. S., **"On-Line, Mass Spectrometric Determination of Ammonia from Oil Shale Pyrolysis Using Isobutane Chemical Ionization,"** *Novel Techniques in Fossil Fuel Mass Spectrometry, ASTM STP 1019,* T. R. Ashe and K. V. Wood, Eds., American Society for Testing and Materials, Philadelphia, 1989, pp. 133–143.

ABSTRACT: Ammonia is the predominant nitrogen-containing species formed during the pyrolysis of oil shale. An on-line method of determining ammonia was developed to study kinetics and chemistry of formation and evolution. While fragmentation of water interferes with electron ionization mass spectrometry, chemical ionization mass spectrometry using isobutane reagent gas gave excellent sensitivity and selectivity for ammonia. All transfer and sampling lines from the retort were heated to approximately 150°C. We successfully measured ammonia evolution from various oil shales retorted under a variety of conditions, including steam retorting.

KEY WORDS: ammonia, mass spectrometry, chemical ionization, oil shale, pyrolysis, retort, on-line methods, real-time methods, determination, steams, fossil fuels

Oil derived from oil shale deposits is a major potential source of liquid fuel and petrochemical products. It is estimated that organic-bearing sedimentary rocks classed as oil shale underlie as much as 20% of the area of the United States [*1*]. The two principal deposits are the Green River Formation located in Colorado, Utah, and Wyoming and the Devonian oil shale deposit that outcrops in Kentucky, Tennessee, Indiana, and Ohio. Although the United States possesses large reserves of oil shale, use of this resource depends heavily on developing an understanding of the chemistry of this material. Workers in this laboratory are in the midst of a long-term project to study the chemistry of oil shale in order to develop economic and environmentally acceptable methods of deriving shale oil.

The most common way of producing oil is to pyrolyze the oil shale in a retort at approximately 500°C. At this temperature, the kerogen (organic material in the oil shale) breaks down to form oil and hydrocarbon gases. In addition, sulfur-, oxygen-, and nitrogen-containing products, both liquid and gaseous, are formed from decomposition and reactions involving both the kerogen and the mineral bearing portions of the shale.

Ammonia is the predominant nitrogen-containing species formed during the pyrolysis of oil shale [*2*]. It is important to study the factors influencing its formation as it could be

[1] Section leader, Organic Analysis Section; fossil fuels chemist; mass spectroscopist, Organic Analyses Section; and chemical engineer, respectively, University of California, Lawrence Livermore National Laboratory, Livermore, CA 94550.

a valuable by-product. In addition, its release is restricted for environmental reasons. One of the most useful tools for studying the chemistry and kinetics of ammonia formation and decomposition is a laboratory scale retort that can be heated using a very precise temperature ramp, coupled with an on-line monitor specific for ammonia. By measuring real-time evolution, kinetic parameters can be calculated for the formation of the ammonia as well as other products. Further, by varying the composition of sweep gas used during retorting, possible reaction pathways can be inferred.

Ammonia, in similar complex mixtures, has been monitored using on-line Fourier transform infrared spectroscopy (FTIR) [*3,4*]. Conventional electron ionization mass spectrometry (EIMS) has been used to monitor the decomposition of NH_4VO_3 [*5*], where spectral interferences such as water (H_2O) were not a problem, as anhydrous conditions were employed and ammonia was 25 to 67% of the amount of the gaseous product formed. The use of chemical ionization (CI) mass spectrometry has been reported for determining ^{15}N-ammonia from biochemical experiments where again ammonia is the major gaseous species [*6,7*]. These measurements were made over a fairly short time span, on the order of several minutes.

There are a number of factors that make mass spectrometry an attractive real-time monitor. It gives results that are generally matrix independent, sensitive, and often specific. The use of EIMS to accurately determine ammonia is hampered by the large amount of water formed in oil shale retorting. Such problems are exacerbated when steam retorting is performed. Water has a large fragment at mass to charge ratio (m/z) of 17 (OH^+) and a smaller but still significant peak at m/z 16 (O^+). In principle, it is possible to remove the contribution of water fragments at m/z 17 by subtracting the contribution from water, based on its cracking pattern from m/z 18. In practice, there is often only a very small amount of ammonia and a large amount of water; thus the noise in the system makes this calculation impractical. In addition, the methane isotope, $^{13}CH_4$, which is present at the level of 1.11% in methane, is a further isobaric interferant at m/z 17. Methane is the major hydrocarbon gas formed during oil shale pyrolysis.

Isobutane chemical ionization readily forms NH_4^+ at m/z 18, while H_3O^+ (m/z 19) is formed with low efficiency, and no measurable $^{13}CH_5^+$ (m/z 18) is formed. This is expected based on the proton affinities of these compounds; CH_4 (131) $<$ H_2O (173) $<$ iC_4H_{10} (197) $<$ NH_3 (205) [*8*]. The protonated water formed may be due, at least partially, to self CI. The amount of H_2O^+ at m/z 18 can be held to $<$1% of the protonated water (m/z 19), and thus causes insignificant interference with the ammonia.

Procedure

Mass Spectrometer

The mass spectrometer used for these studies is a triple quadrupole mass spectrometer (TQMS) built at Lawrence Livermore National Laboratory [*9*], equipped with an HP 5985B gas chromatography/mass spectrometry (GC/MS) EI-CI source. Although a normal, nontandem mass spectrometer with a CI source would have been adequate, the TQMS is used here because it is being used as a generic monitor of fossil fuel chemistry, including the determination of trace sulfur-containing gases [*10,11*]. Since ammonia intensity is influenced by the CI gas pressure, it is critical to hold the reagent gas pressure constant. Source CI gas pressure was monitored with a Granville-Phillips Convectron 275 pressure meter connected to the ~36-cm (14 in.) long by 6.4-mm (¼ in.) outside diameter (OD) stainless steel (SS) tubing leading into the source. The analog output from this meter feeds the input of a Datametrics Model 1404 valve controller. This controller is connected to a Datametrics Model 1610-A electronic valve that is mounted in the reagent gas line and controls the pressure of isobutane at 600 $\pm$ 0.5 mtorr as read on the gage. The true pressure is $<$30 mtorr, calculated by ratioing the sensitivity of nitrogen (used to calibrate the gage) and isobutane.

The inlet to the mass spectrometer utilizes a Balzers Model GES 010, which consists of a 0.3-mm ID SS capillary, 100 cm long that leads into a split valve, where a small amount of gas enters the mass spectrometer through a frit. The remainder of the gas exiting the capillary is pumped away using a mechanical pump. This system is designed to eliminate fractionation of the gas based on molecular weight. The capillary and splitter valve are heated to 200°C, and the line from the splitter valve to the source is heated to over 100°C. The flow of gas through the capillary is very precise at about 15 cm^3/min.

Retort

The laboratory-scale retort is designed to heat 25 or 60 g of oil shale at a precise and uniform rate from ambient temperature to over 900°C. The outer (main) furnace is 3.8 cm ID, 15 cm OD, and 28 cm high and uses nonlinear heating coil spacing to achieve uniform temperature throughout the length. The SS retorts used in these experiments are 4.4 cm high and fabricated using a pair of concentric cylinders, such that an inner rod heater, 1.6 cm OD, can be installed to provide radial uniformity of the shale temperature ($\pm$4°C at a heating rate of 4°C/min). The temperature of the shale is monitored by a sheathed thermocouple that is in direct contact with the specimen. In addition, control thermocouples are mounted between the retort and the heaters. The inner heater and outer tube furnace have independent Lindberg Model 59344 controllers, but they are driven by the same temperature ramp program.

On occasion, silica glass (quartz) retorts are used. In this case, no inner heater is used. A sheathed thermocouple is sealed into a fitting on top (outside the furnace) of the quartz retort and is buried in the shale at the specimen midpoint.

The temperature of the retort is monitored by a Hewlett Packard (HP) 85 computer using an HP 3497A thermocouple conditioning and digitizing system which communicates over a Hewlett Packard Interface Buss (HPIB). The thermocouples are read every minute and stored on a tape cartridge. The HP 85 computer is used to control the heating rate by updating temperature settings of the controllers every 12 s. Oil shale temperature as well as sweep gas flow rate and pressure is stored, along with the mass intensity data, on the TQMS data base system. A DEC LSI 11/73 computer, running the RSX 11M+ operating system, was used with this data base [*12*].

To assure leak-free operation, the SS retort is welded shut after the shale is placed in it. A sweep gas is brought into the retort for all experiments. The sweep gas is often pure argon or argon and steam, but occasionally gases such as hydrogen or nitrogen are used to probe specific chemical reactions. The flow of the sweep gas (except for steam) is controlled by a pneumatic flow controller (Brooks 8744) that maintains the flow at a set point with a precision of $\pm$0.25%. The flow is monitored using both a rotameter as well as an electronic flow monitor, Datametrics Model 831-500C4. The pressure of the sweep gas is recorded by a Teledyne 2415 pressure transducer. Steam is generated by forcing water, using 241 kPa (35 psig) argon, into coils of 3.2 mm (⅛ in.) SS tubing in a furnace at ~200°C. The flow was controlled by a heated needle valve at the steam furnace exit.

The gas exiting the retort is maintained at a temperature of $\geq$140°C to eliminate condensation of water (ammonia will dissolve in liquid water as a buffered solution of $(NH_4)_2CO_3$ or $(NH_4)_2S$). A trap, also at about 140°C, is located down stream of the retort to collect the heavy oil produced. Gas that passes the end of the TQMS inlet capillary is exhausted into a fume hood.

Samples and Standards

The Green River Formation (western) oil shale came from the Anvil Points mine in Colorado. The Devonian (eastern) oil shale came from the Clegg Creek Member of the New Albany shale in Bullit County, Kentucky. Specimens were removed from large blocks of shale, ground and sieved to obtain a −6/+50 mesh size range, and stored in brown bottles under argon. Buddingtonite ($NH_4AlSi_3O_8 \cdot ½H_2O$ [*13*]), an inorganic source of nitrogen in western shales, was obtained as a museum-grade specimen from the Smithsonian Institution.

Ammonia calibration standards are made up ~1% in argon and are Matheson Analyzed grade. The isobutane CI reagent gas is Airco Instrument grade, and is passed through an Alltech Model 8128 gas purifier to remove water.

Experimental Work and Results

Initial experiments used ion-molecule reactions in the collision cell of the triple quadrupole mass spectrometer to remove the water interference using EIMS. At high pressures (1+ mtorr) of nitric oxide (NO) in the collision cell of the TQMS, charge exchange occurred with H_2O^+ significantly more than with NH_4^+ (preferential charge exchange), as is shown in Fig. 1. Thus, much more of the ammonium ion was detected at *m/z* 18 than water ion. Further, it was found that scattering due to the collision gas is proportional to the log of the pressure (analogous to absorption of light as described by Beer's law). The difference in the slope of the two lines seen in Fig. 1 is due to charge exchange; the upper line (NH_4^+) represents scattering only, while the lower line (H_2O^+) represents the combined effects of scattering and charge exchange.

Although charge exchange cleanup is useful in some circumstances, in general, we have been able to achieve sufficient selectivity using favorable CI conditions. An important requirement is that as much water as possible be removed from the CI reagent gas. This is achieved by drying the isobutane shortly before it enters the source, and heating all the

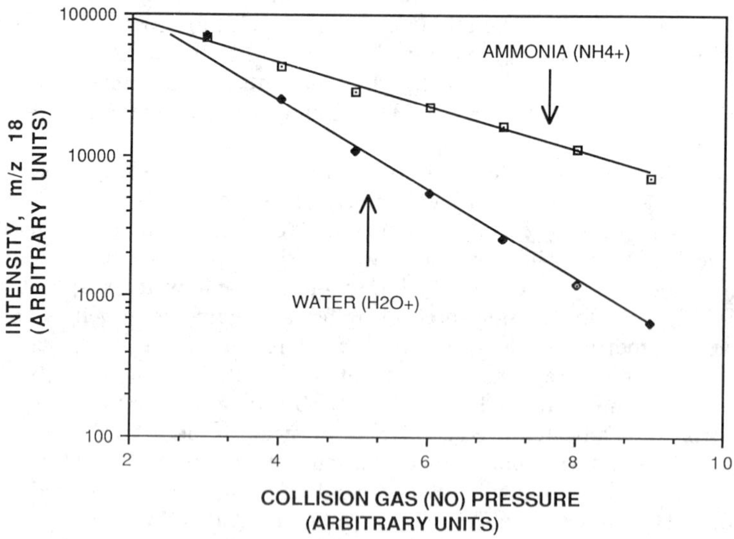

FIG. 1—*The use of nitric oxide (NO) as the collision gas in the TQMS shows preferential charge exchange with water ions.*

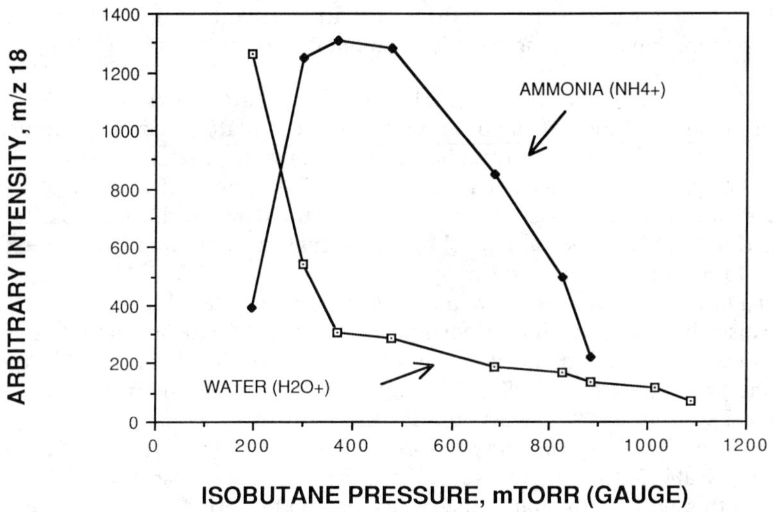

FIG. 2—*The pressure of isobutane is critical to maximize ammonia intensity, minimize water intensity.*

lines from the dryer to the source. The most important modification of CI conditions for reducing water interference is increasing the CI reagent gas pressure in the source such that H_2O^+/H_3O^+ is <0.01. Figure 2 shows the effect of reagent gas pressure on ammonia and water intensity. The increased pressure of isobutane increases the proton transfer/electron ionization ratio. Combining these techniques allows us to determine ammonia at <50 ppm in steam, as can be seen in Fig. 3 where a pulse of 100 ppm of ammonia is added to a flow of steam.

Many of our experiments involved heating shale at 4°C/min from room temperature to 900°C. This means taking data continuously for a period of almost 4 h. The TQMS is a

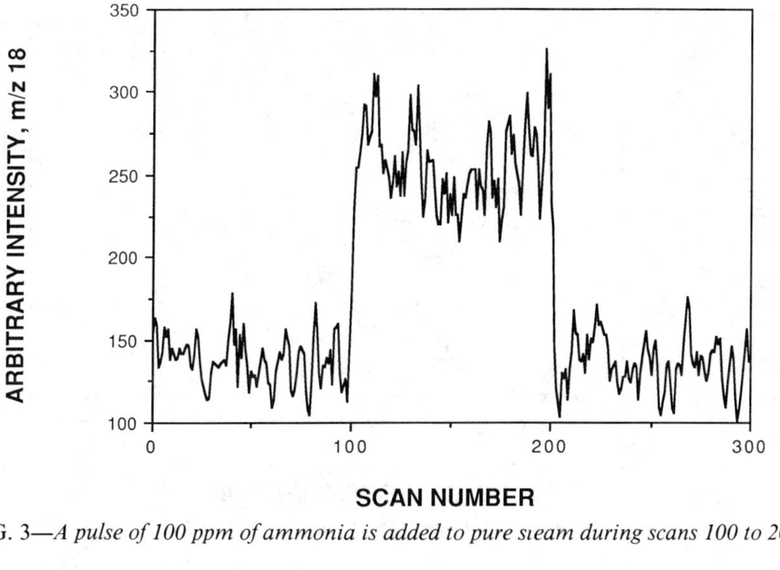

FIG. 3—*A pulse of 100 ppm of ammonia is added to pure steam during scans 100 to 200.*

qualitative instrument, although it can do quantitative work if calibrated frequently. There are several factors that affect long-term instrumental stability.

1. *Secondary electron multiplier (SEM) gain*—Electron currents of ~1 μA or larger, appearing at the anode of the discrete dynode secondary electron multiplier (SEM), cause up to a 15% loss in multiplier gain. To prevent problems of this nature, the present data have been acquired using selected-ion monitoring such that the intense peaks from isobutane fragments do not strike the detector. In addition, the voltage applied to the SEM is set at a level to ensure that the ammonia signals never exceed 1 μA of multiplier anode current.
2. *Changes in transmission of the quadrupole analyzers*—Rapid scanning or peak jumping causes localized heating in portions of the radio frequency circuits that may in turn alter the resolution and thus the transmission of the quadrupole analyzers [*17*]. For the present work, this effect is minimized by running the instrument under the identical scan mode for at least 30 min before real data are acquired.
3. *Changes in CI reagent gas pressure*—As seen in Fig. 2, the intensity of the ammonia is strongly affected by the pressure of the isobutane. As described in the Procedure section, this pressure is maintained constant (600 mtorr ± 0.5 mtorr gage reading) using a servo feedback system.

Noise is further reduced by repeated scanning and averaging. The analog to digital converter runs at a rate of 10 μs per conversion; each scan entails 10 000 conversions at each mass, requiring 100 ms. Seven different masses are monitored for a total of 700 ms/scan. Each acquisition consists of 30 s of data; thus 43 scans are averaged. Any stored point is made up of 430 000 individual readings per mass, reducing white noise by a factor of ~650 over that from a single conversion.

Fragments of iosbutane are also monitored to help eliminate short- and long-term drift, with the pressure of isobutane held constant to <1%. Four isobutane fragments (at m/z 13, 25, 37, and 59) are scanned and stored. These are chosen to be near masses of interest and to be of a similar intensity to ammonia. Figure 4*a* shows the raw data for ammonia

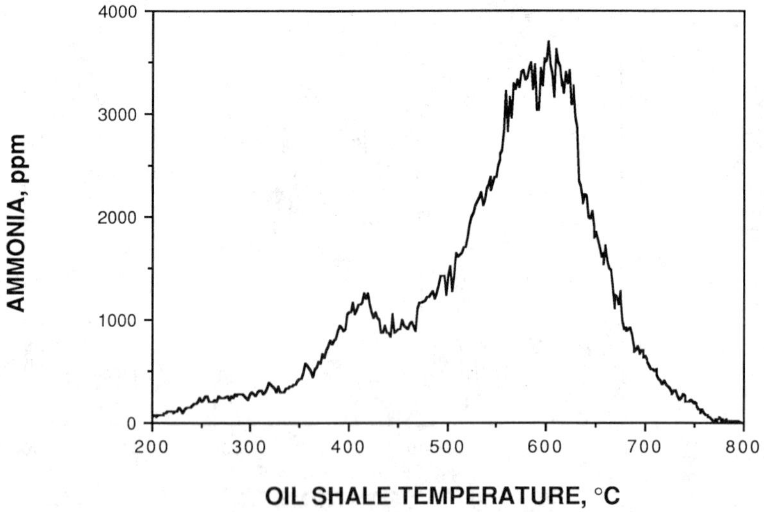

FIG. 4*a*—*Raw concentration data for ammonia from the western shale.*

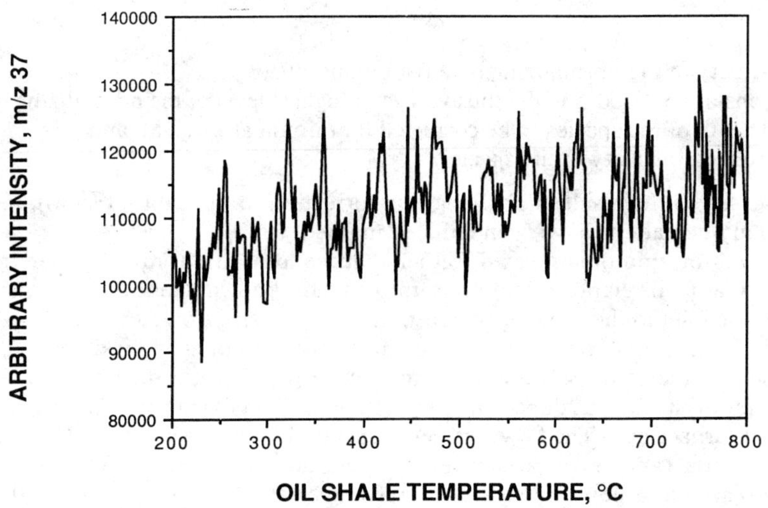

FIG. 4*b*—*Intensity of a fragment from isobutane, used to correct for instrumental variations.*

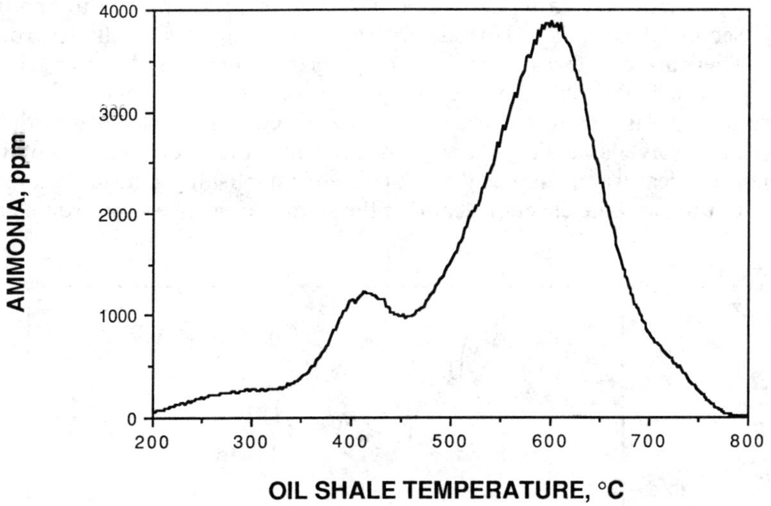

FIG. 4*c*—*Concentration data for ammonia from the western shale after correcting for instrumental variations (4b).*

evolution from oil shale. Note the short-term intensity change in this figure. Figure 4*b* is the *m*/*z* 37 fragment from isobutane monitored in this same experiment. Again note both the short-term intensity change and the long-term drift in this data. The intensity of the isobutane peak at *m*/*z* 37 may be used to correct the observed ammonia data on a point by point basis. Specifically

$$(I_N/C) \times I_s = I_c$$

where

I_N = intensity of the normalizing data (isobutane *m/z* 37);
C = a constant entered, usually the average or highest level of the normalizing intensity;
I_s = intensity of the species to be corrected (ammonia at *m/z* 18); and
I_c = corrected intensity of the species.

Figure 4*c* shows the result of applying this correction to the data set shown in Fig. 4*a*. Instrumental fluctuations have been reduced markedly.

The volume of ammonia created, rather than concentration, must be determined to allow meaningful correlations with other information known about the shales. Since the pressure at the inlet to the capillary is constant, the mass spectrometer measures only concentration. Therefore, measurement of the amount of ammonia evolved demands knowledge of the volume of the gas due to formation of pyrolysis products. In initial experiments, the pyrolysis gas was doped with a constant amount (4 cm^3/min) of trimethylamine at the retort furnace exit. The flow was held constant by immersing the $(CH_3)_3N$ tank and 5 m of 0.13 mm capillary in a constant temperature water bath at 37.2°C. The trimethylamine served as a flow and sensitivity reference. The M + 1 ion at *m/z* 60, $(CH_3)_3N$ + H, is monitored and the decreases in intensity are inversely proportional to the increases in total gas flow. The ammonia data are then corrected using this data to give absolute volume information.

Several problems exist with the use of trimethylamine as an internal standard. First, this material has properties and a repugnant odor that make it a nuisance to handle in the laboratory. Secondly, it tends to corrode critical systems such as capillaries, rotameters, and valves. There are very few appropriate materials that can be handled as a gas, that give a mass not isobaric with isobutane, and that are accessible by isobutane CI.

In later experiments steam was used as a flow tracer. During steam retorting experiments, it is already available. For other systems, it is introduced downstream of the retort. Steam is not an ideal tracer, due to the difficulty of establishing a steady steam flow, its low proton affinity, and the effect of steam on the source of the mass spectrometer. Never-

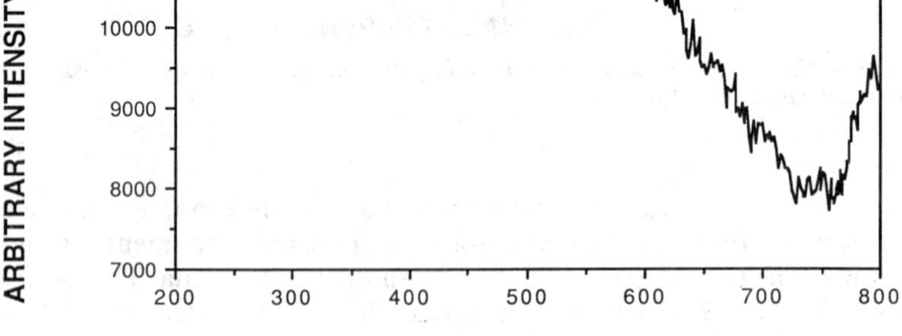

FIG. 5—*Intensity of the hydronium ion (H_3O^+) from the steam tracer used to measure total flow variation due to the evolution of gases from the pyrolysis of oil shale.*

theless, steam can successfully be used as a tracer. Figure 5 shows the intensity measured at m/z 19 for steam from the pyrolysis of oil shale. The dip at about 450°C is due to oil and pyrolysis gas evolution. The dip at about 750°C is due mainly to carbon dioxide evolution from carbonate mineral decomposition. This data can be applied as a correction factor to the ammonia, allowing results to be reported as cubic centimetres of ammonia per minute per gram of shale.

Discussion

The application of this technique to oil shale has been discussed in several reports [*14,15*]. In addition, it has been applied to Asphalt Ridge tar sand. Initial experiments studied ammonia decomposition in a stainless steel retort using steam, hydrogen (H_2), or inert sweep gas. The temperature at which 50% of the ammonia decomposed was 700°C with argon, 800°C with hydrogen, and close to 1000°C with 56% steam in argon.

In western but not eastern oil shale, a portion of the nitrogen is inorganic, in the form of the mineral Buddingtonite [*16*]. The formation of ammonia from this mineral has been studied and almost three times as much NH_3 is evolved with steam retorting as with argon. The temperature of maximum ammonia evolution was 630°C when the heating rate was 4°C/min (Fig. 6).

Finally, ammonia evolution profiles from steam pyrolysis of western and eastern oil shales are compared in Figs. 7 and 8. The peak at 450°C is concurrent with oil production. The peak at 720 to 750°C is a combination of Buddingtonite decomposition (for the western oil shale) and reactions of organic nitrogen sources with hydrogen donors during charring. Partial plugging of the retort occurred with the eastern oil shale, which reduces the accuracy of the determination of ammonia.

Acknowledgments

The authors wish to thank Art Lewis, project leader for Fossil Fuels for his continued support. Further thanks are extended Michael May, associate director in charge of the

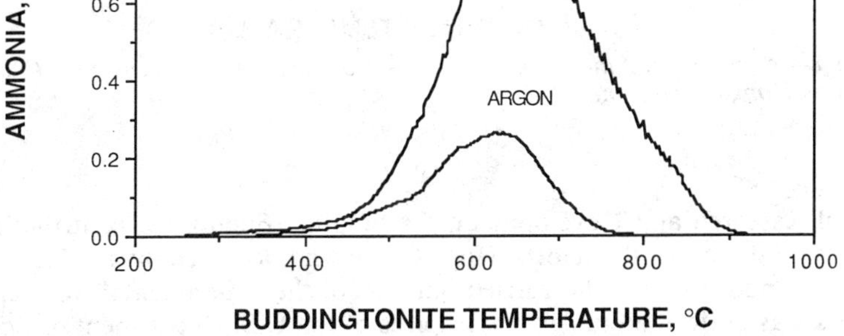

FIG. 6—*Pyrolysis of Buddingtonite shows that almost three times as much ammonia survives when steam sweep is used.*

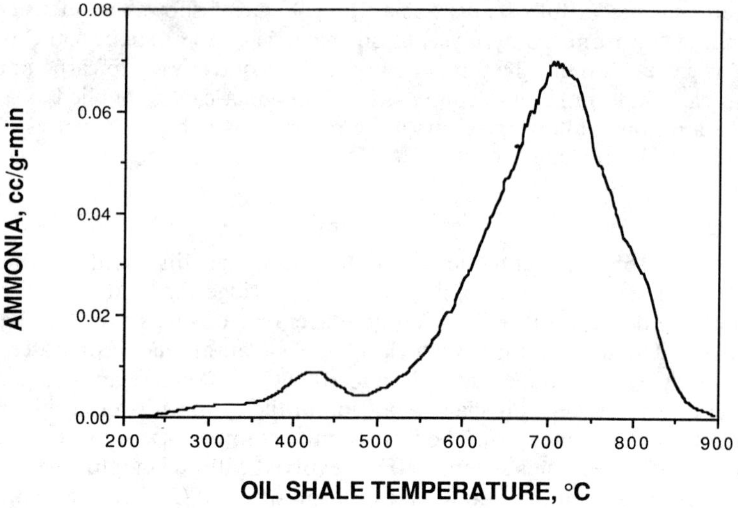

FIG. 7—*Ammonia evolution profile from the steam pyrolysis of a western oil shale, Anvil Points 100 L/mg (24 gal/ton).*

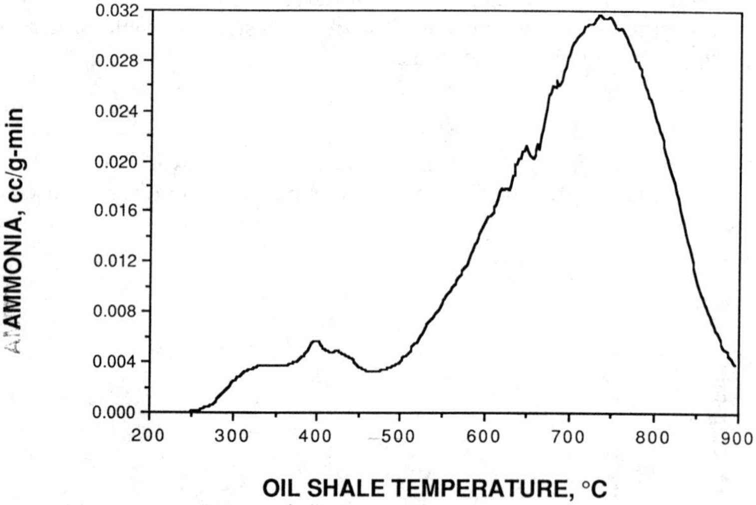

FIG. 8—*Ammonia evolution profile from the steam pyrolysis of an eastern oil shale, New Albany 54 L/mg (13 gal/ton).*

Institutional Research and Development, for funding. Finally the contributions of Ken Foster who set up most of the retorts, Hugh Gregg who provided invaluable software support, and Armando Alcaraz who carried out part of the experimental work are acknowledged. Work was performed under the auspices of the U.S. Department of Energy by the Lawrence Livermore National Laboratory under Contract W-7405-ENG-48.

References

[*1*] Miknis, F. P. and Smith, J. W., "An NMR Survey of United States Oil Shales," *Proceedings,* Fifteenth Oil Shale Symposium, J. H. Gary, Ed., Colorado School of Mines Press, Golden, CO, Aug. 1982, pp. 50–62.

[*2*] Fruchter, J. S., Wilkerson, C. L., Sklarew, D. S., Olsen, K. B., and Ondov, J. M., "Potential Air Emissions from Oil Shale Retorting," *Oil Shale: the Environmental Challenges III,* K. K. Petersen, Ed., Colorado School of Mines Press, Golden, CO, 1983, pp. 139–164.

[*3*] Solomon, P. R. and Hamblen, D. G., "Pyrolysis," *Chemistry of Coal Conversion,* R. H. Schlosberg, Ed., Plenum Press, New York, 1985, pp. 138–145.

[*4*] Whelan, J. K., Solomon, P. R., Deshpande, G. V., and Carangelo, R. M., *Energy and Fuels,* Vol. 2, No. 1, Jan. 1988, pp. 65–73.

[*5*] Dziembaj, R., "Mass-Spectrometric Study of the Thermal Decomposition of NH_4VO_3," *Geterogennyi Katal, (Heterogeneous Catalysis),* D. Shopov, A. Andreev, and A. Palazov, Eds., Sofia, Bulgaria, 1975, pp. 461–467.

[*6*] Lundeen, C. V., Viscomi, A. S., and Field, F. H., *Analytical Chemistry,* Vol. 45, No. 7, June 1973, pp. 1288–1290.

[*7*] Ukida, M. and Shimada, Y., *Gastroenterologia Japonica,* Vol. 14, No. 2, April 1979, pp. 110–113.

[*8*] Harrison, A. G., *Chemical Ionization Mass Spectrometry,* CRC Press, FL, 1983, pp. 31–35.

[*9*] Wong, C. M., Crawford, R. W., Barton, V. C., Brand, H. R., Neufeld, K. W., and Bowman, J. E., *Review of Scientific Instruments,* Vol. 54, No. 8, Aug. 1983, pp. 996–1004.

[*10*] Wong, C. M., Crawford, R. W., and Burnham, A. K., *Analytical Chemistry,* Vol. 56, No. 5, March 1984, pp. 390–395.

[*11*] Wong, C. M., Crawford, R. W., and Yost, R. A., "An Evaluation of Triple Quadrupole Mass Spectrometry/Mass Spectrometry (MS/MS) for On-Line Gas Analysis of Trace Sulfur Compounds from Oil Shale Processing," *Mass Spectrometric Characterization of Shale Oils, ASTM STP 902,* T. Aczel, Ed., American Society for Testing and Materials, Philadelphia, 1986, pp. 106–120.

[*12*] Crawford, R. W., Brand, H. R., Wong, C. M., Gregg, H. R., Hoffman, P. A., and Enke, C. G., *Analytical Chemistry,* Vol. 56, No. 7, June 1984, pp. 1121–1127.

[*13*] Erd, R. C., White, D. E., Fahey, J. J., and Lee, D. E., *American Mineralogist,* Vol. 49, Nos. 7–8, 1964, pp. 863–879.

[*14*] Coburn, T. T., Crawford, R. W., Gregg, H. R., and Oh, M. S., "Reactive Gases Evolved During Pyrolysis of Devonian Oil Shale," *Proceedings,* 1986 Eastern Oil Shale Symposium, Lexington, KY, 19–21 Nov. 1986, pp. 291–299.

[*15*] Oh, M. S., Taylor, R. W., Coburn, T. T., and Crawford, R. W., *Energy and Fuels,* Vol. 2, No. 1, Jan. 1988, pp. 100–105.

[*16*] Taylor, R. W., Burnham, A. K., Smith, G. S., Sanborn, R. H., and Gregory, L. S. in *Proceedings,* Division of Fuel Chemistry, American Chemical Society, Vol. 30, No. 3, pp. 338–348.

[*17*] Miller, P. E. and Denton, M. B., *Journal of Chemical Education,* Vol. 63, July 1986, pp. 617–622.

Koli Taghizadeh,[1] Burtron H. Davis,[1] Willem Windig,[2] and Henk L. C. Meuzelaar[3]

Off-Line Process Monitoring of Coal-Derived Liquid Fuels by Computer-Assisted Low-Voltage Mass Spectrometry

REFERENCE: Taghizadeh, K., Davis, B. H., Windig, W., and Meuzelaar, H. L. C., **"Off-Line Process Monitoring of Coal-Derived Liquid Fuels by Computer-Assisted Low-Voltage Mass Spectrometry,"** *Novel Techniques in Fossil Fuel Mass Spectrometry, ASTM STP 1019,* T. R. Ashe and K. V. Wood, Eds., American Society for Testing and Materials, Philadelphia, 1989, pp. 144–158.

ABSTRACT: Compositional changes of Coal-Derived Liquids (CDLs) during hydrotreatment on fixed bed catalysts can be investigated by low-voltage (12 eV) mass spectrometry (MS) in combination with multivariate analysis without chromatographic pre-separation. Low-voltage (12 eV) MS data were obtained and compared under two different inlet conditions; (1) MS inlet at ambient temperatures and (2) with the MS inlet preheated to about 200°C. The major trend found in both data sets by Factor analysis was the hydrotreatment effect, namely, hydroaromatic (hydrotreater product) versus aromatic (hydrotreater feed) compounds. A second "time + temperature" trend that reflected the aging of the catalyst as well as the temperature of the hydrotreater reactor was only obtained under ambient inlet temperature conditions. Numerically extracted spectra along the second trend show that more of the low molecular weight components were found early in the process and also at lower reactor temperatures; whereas more of the higher molecular weight components, primarily alkyl substituted polynuclear aromatics, were obtained later in the process and at higher temperatures of the hydrotreater.

KEY WORDS: fossil fuels, mass spectrometry, coal-derived liquids, low voltage mass spectrometry, multivariate analysis, hydrotreatment, catalyst aging, aromatics, hydroaromatics, polynuclear aromatics

Investigating the effects of hydrotreatment or other process related procedures and conditions on the composition of coal derived liquids (CDLs) such as obtained from the Wilsonville ITSL (Integrated Two-Stage Liquefaction) process requires powerful analytical techniques. Ideally, coal liquefaction process monitoring techniques should enable differentiation and identification (or at least classification) of hundreds of CDL components while detecting relatively small changes in the relative concentration of any or all of these

[1] Research associate and program director, respectively, Kentucky Energy Cabinet Laboratory, Lexington, KY 40512.

[2] Senior analytical chemist, Kodak, Rochester, NY 14650.

[3] Professor of Materials Science and Engineering University of Utah, Center for Microanalysis and Reaction Chemistry, Salt Lake City, UT 84112.

components. Moreover, in view of the desirability to analyze many separate CDL aliquots obtained during the course of a single run, successful candidate techniques should be relatively fast.

Mass spectrometry (MS) has long been recognized as one of the most powerful analytical techniques for complex organic mixtures. Especially when combined with chromatographic pre-separation methods such as gas chromatography (GC/MS) or liquid chromatography (LC/MS), hundreds of individual components can be differentiated and identified in a single sample. Unfortunately, GC/MS and LC/MS methods are quite time-consuming unless used in a "selective ion monitoring" mode, for example, when monitoring known components or trends. Direct mass spectrometric analysis, however, although extremely fast, usually provides insufficient information to allow satisfactory chemical characterization of complex CDLs. This is partly due to the severe fragmentation occurring during electron ionization (EI) at 70 eV, the most widely used ionization practice in MS, which limits the maximum number of components that can be simultaneously detected and analyzed due to overlapping fragment ion signals.

Consequently, low-voltage (20 eV) EI-MS has been widely used to reduce fragmentation formation. Moreover, lowering the ionization energy further, for example, down to 12 eV level used in this study, markedly reduces the residual fragmentation processes [*1*]. Nevertheless, "soft" ionization methods such as field ionization (FI) or chemical ionization (CI) are more successful in reducing fragmentation tendencies. The FI-MS techniques have proven to be capable of producing highly characteristic molecular profiles of CDLs [*2–6*] and other liquid fossil fuels. However, even when containing exclusively molecular ions, a nominal resolution mass spectrum does not differentiate between ion species of the same nominal m/z ratio. Consequently, high resolution MS techniques have been used to obtain additional separation, for example, between hydrocarbon compound series and heteroatomic compounds [*7*].

Since computer-assisted analysis of spectral time series has proven capable of detecting and monitoring relatively minor components in complex mixtures [*8*], the authors decided to use this approach for analyzing a series of CDLs from the Wilsonville, Alabama, pilot plant during a three-month run of Illinois No. 6 coal (Burning Star Mine) while operating in the ITSL mode. Besides testing the applicability of computer assisted time-series MS techniques, the purpose of this study was to investigate the effects of hydrotreatment as well as time- and temperature-dependent variations in process conditions on the composition of the Wilsonville coal liquids.

Experimental

A schematic flow diagram of the Wilsonville advanced coal liquefaction facility in the Integrated Two-Stage Liquefaction (ITSL) mode is shown in Fig. 1. Briefly, the feed to the second stage hydrotreater is a mixture of thermal distillate (450°F+) and critical solvent de-ashed thermal residue (CSD-TR). The hydrotreater "atmospheric flashed bottoms" stream is recycled to the first stage dissolver. The hydrotreater temperature was increased during the course of the operation; more detailed operational information can be found elsewhere [*9–10*].

The data presented here were generated for the CDL samples collected during run No. 244 of Illinois No. 6 coal from the Burning Star Mine. The sampling points for both hydrotreater feed and hydrotreater atmospheric flashed bottoms samples (for simplicity to be referred to as "hydrotreater product") are shown in Fig. 1, whereas the various samples analyzed are identified in Table 1.

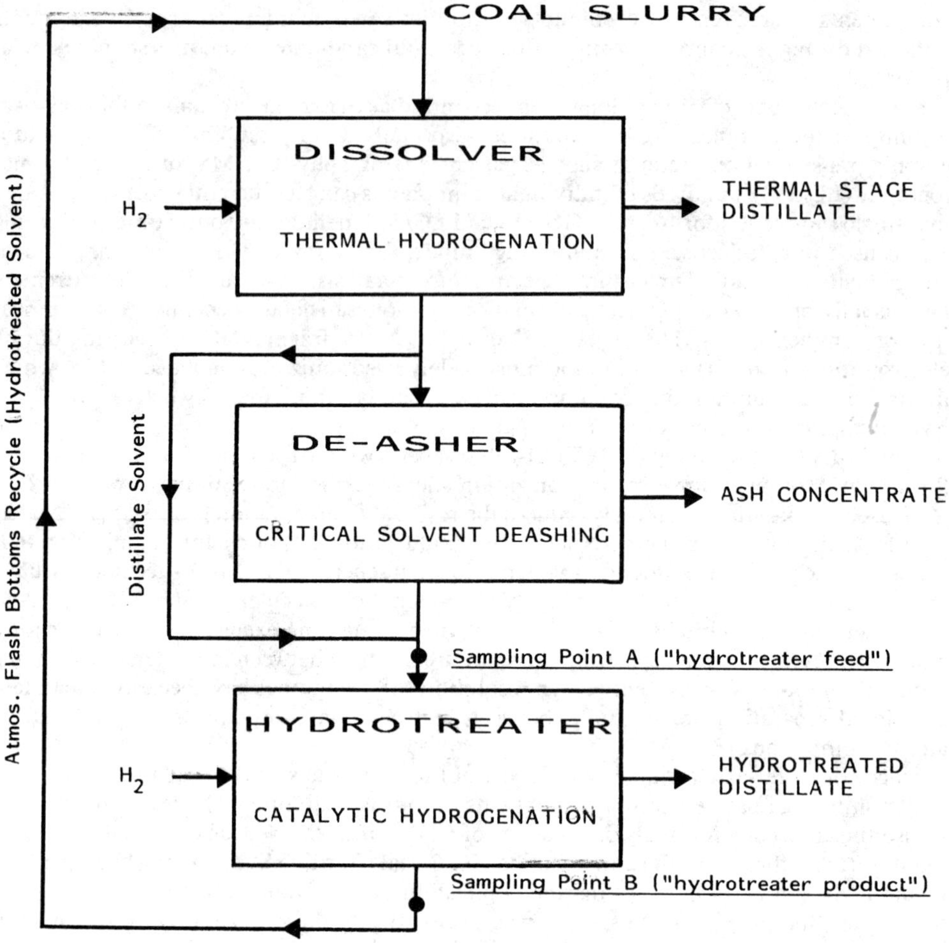

FIG. 1—*Schematic of Wilsonville Integrated Two Stage Liquefaction (ITSL) process.*

TABLE 1—*Sample identification.*

Sampling Dates	Hydrotreater Feed Sample No.	Hydrotreater Product Sample No.
7-06-83[a]	1	11
7-10-83	2	12
7-15-83	3	13
7-21-83	4	14
7-24-83[a]	5	15
8-07-83[a]	6	16
8-17-83	7	17
8-28-83	8	18
8-30-83	9	19
9-08-83[a]	10	20

[a] Samples analyzed under the preheated inlet conditions.

Sample Preparation

About 2 mg of each CDL were weighed and dissolved in 1 mL of a 1:1 benzene:methanol mixture (in some cases an ultrasonic bath had to be used to dissolve the CDL completely). Solutions were stored at −10°C prior to analysis.

Low-Voltage Mass Spectrometry

Low-voltage MS analyses were carried out on the Extranuclear 5000-1 quadrupole system with Curie-point heating inlet shown in Fig. 2. A detailed description of the system can be found elsewhere [*11*]. Samples were introduced under two different conditions: (1) MS inlet at ambient temperature and (2) MS inlet preheated to about 200°C.

Operating conditions for the ambient inlet experiments were as follows: sample size, 10 μg; temperature-rise time, 5.5 to 6.0 s; equilibrium temperature, 610°C; total heating time, 10 s; electron energy (set value), 12 eV; scanning rate, 1000 amu/s; mass range, *m/z* 20 to 260; and total scanning time, 20 s. Operating conditions for the preheated inlet experiments were the same as for the ambient inlet conditions except for the mass range of *m/z* 100 to 400 and total scanning time of 2 min.

All spectra were recorded by computer and summed to obtain a single, time-integrated spectrum. Samples introduced at ambient inlet temperature were analyzed in triplicate or in quadruplicate, but samples introduced at 200°C were analyzed only once to minimize ion source contamination.

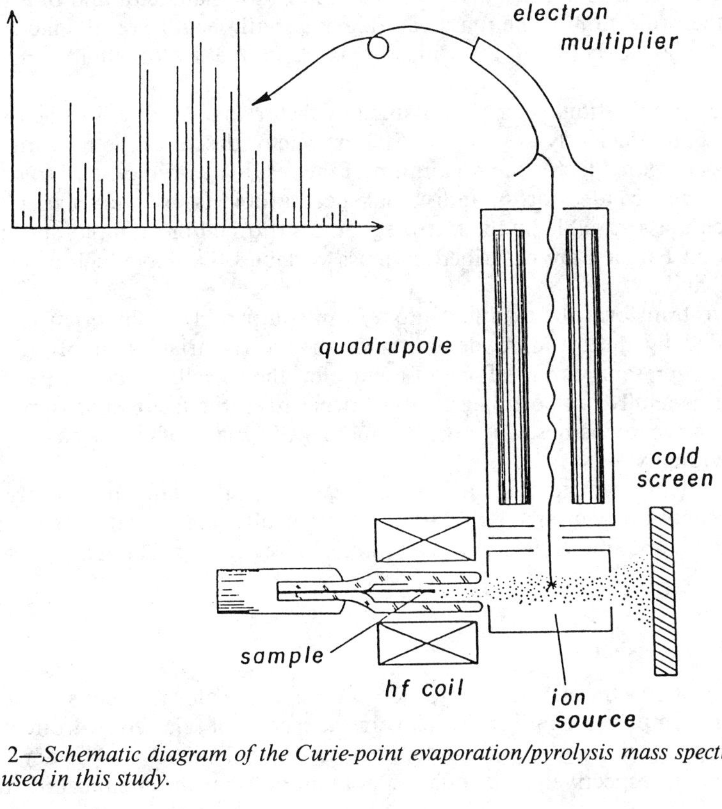

FIG. 2—*Schematic diagram of the Curie-point evaporation/pyrolysis mass spectrometry system used in this study.*

Data Analysis

All data analysis was carried out using the SIGMA (System for Interactive Graphic-oriented Multivariate Analysis) Program [*12*] developed at the University of Utah, Biomaterials Profiling Center. Prior to multivariate analysis, spectra were normalized with regard to total ion intensity in order to correct for sample size and maximize the overall reproducibility of the replicate spectra by excluding from the normalization process those peaks or spectra with unusually large variance contributions in order to minimize error propagation over the data set.

Since in many data analysis problems most interest is focused on relatively minor differences between samples, variables are often transformed prior to data analysis by subtracting the mean. This transformation is generally accompanied by a division of the variables by their standard deviation, in order to give all variables an equal weight in the data analysis. This transformation is known as "standardization" or "autoscaling." It will be clear that the result of this transformation is that all variables now have a mean value of zero and a standard deviation of one.

The primary objective of factor analysis is reduction of the data by redundancy in the data set. Factors are calculated according to the following equation

$$L_j = \alpha_{1j}m_1 + \alpha_{2j}m_2 + \ldots \alpha_{nj}m_n \; j = 1,2, \ldots m \; (m \leq n) \qquad (1)$$

where L_j is the linear combination of the n mass variables. The coefficients, α_{ij}, are the loadings of the variables, m_i, describing the partial contribution of variables, m_i to L_j. The loadings are correlation coefficients of the variables with that factor and thus may be considered as the projection of the mass variable axis, with length l on the factor, where the cosine of the angle between the "mass axis" and the factor equals the correlation coefficient.

The linear combinations (factors) obtained by factor analysis describe the total variance in the data set as efficiently as possible. The first factor describes the maximum variance, the second orthogonal factor the maximum of the residual variance, and so on.

"Scores" are calculated for the individual spectra (objects) by multiplying the intensity value of each mass variable by the factor equations (Eq 1). For example, spectra with positive scores on Factor I are described by mass variables that have high positive loadings on Factor I, and so on.

In order to transform the loadings into a form comparable to the original spectra, they are multiplied by the standard deviation of the mass variables involved. The values obtained are represented in the form of a bar plot, the so-called "factor spectrum." Since the loadings generally can be either positive or negative, the factor spectrum exhibits positive and negative intensities. The positive and negative parts of the spectrum can be interpreted individually.

Subsequently, rotated factor solutions are sought that exhibit maximum correlation with underlying chemical components or trends, thus enabling a definite degree of chemical interpretation sometimes referred to as "numerical extraction of chemical components" [*13–17*].

Results and Discussion

The averaged spectra of hydrotreater feed and hydrotreater product samples (Data Set 1) at ambient temperature and also the averaged spectra of selected hydrotreater feed and hydrotreater product samples in the inlet preheated to about 200°C (Data Set 2) are shown in Figs. 3 and 4, respectively. It should be pointed out that the chemical identities of the

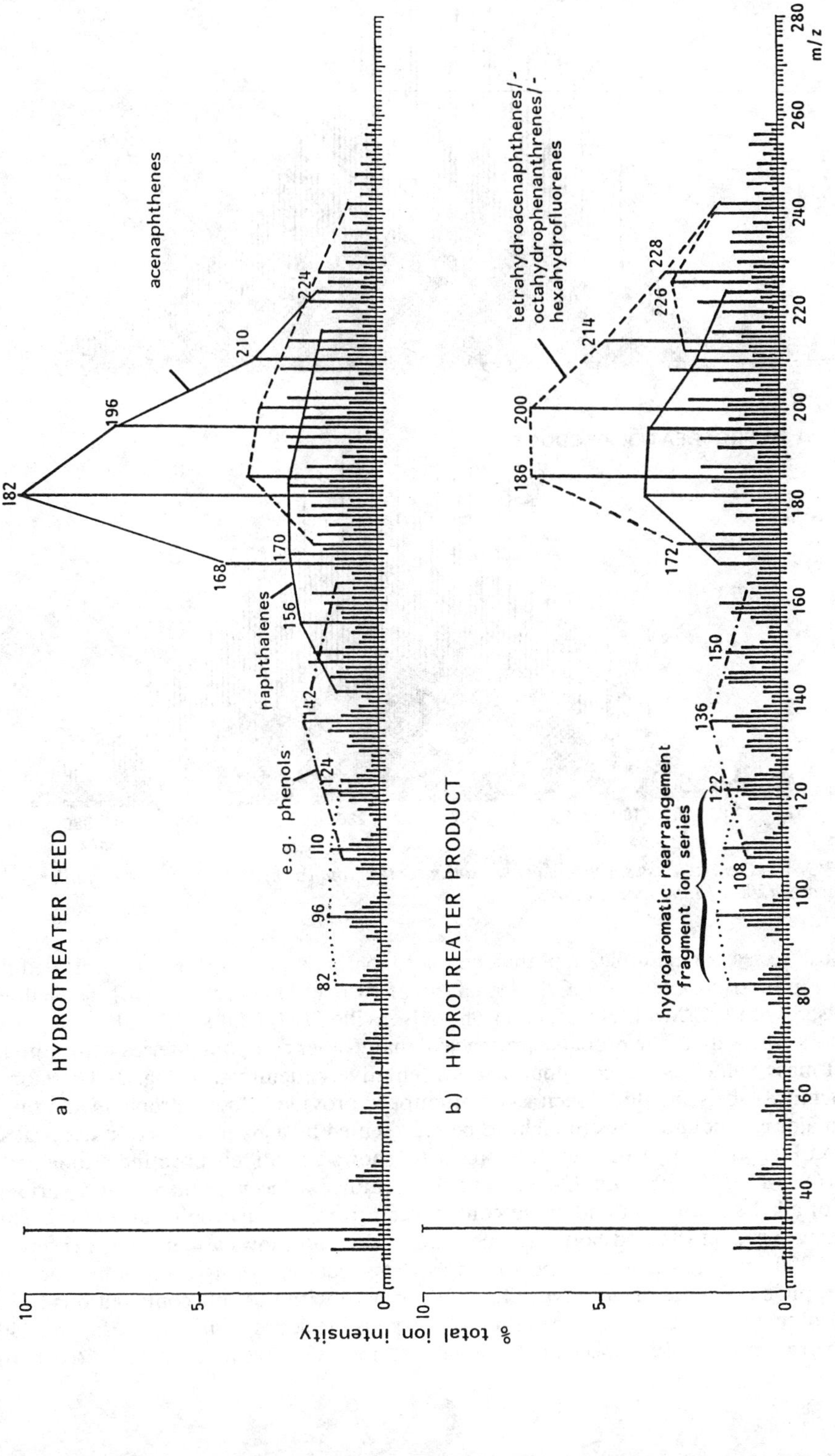

FIG. 3—*Average mass spectra of* (a) *hydrotreater feed and* (b) *hydrotreater product under ambient inlet temperatures.*

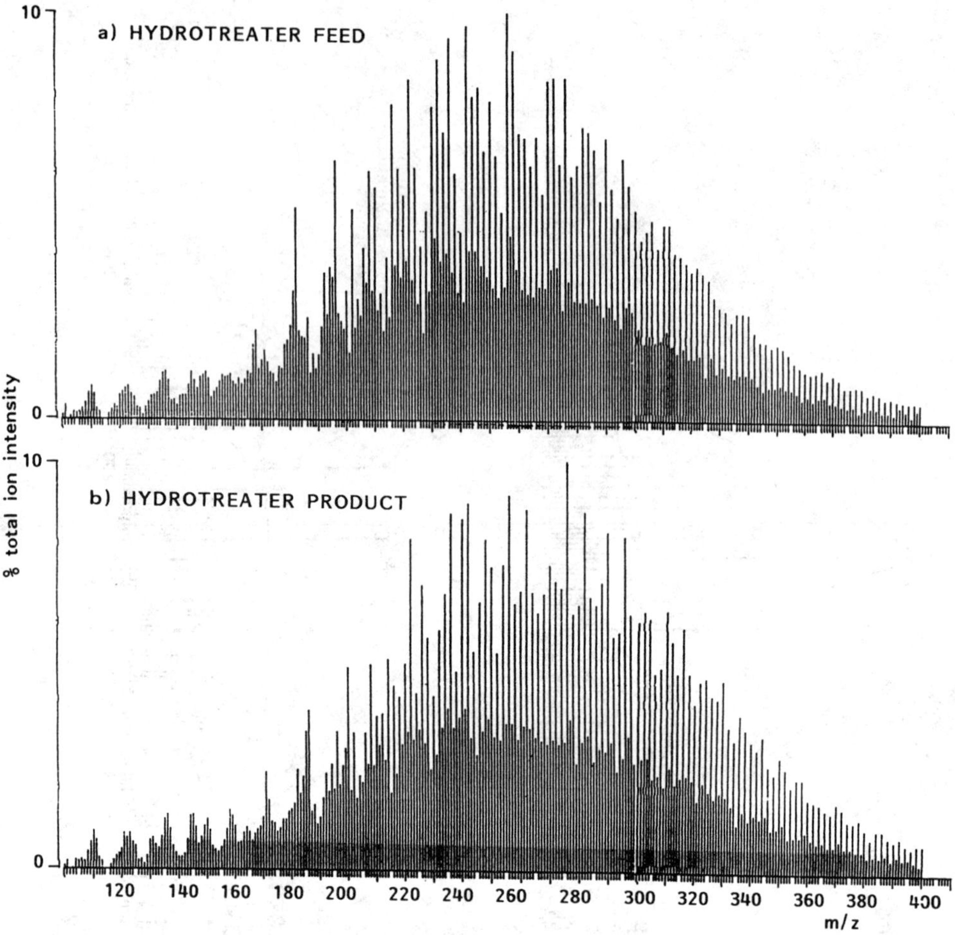

FIG. 4—*Average mass spectra of* (a) *hydrotreater feed and* (b) *hydrotreater product under preheated (200°C) inlet temperatures.*

major peak series, for example, naphthalenes, acenaphthenes, phenols, etc. as indicated in Fig. 3 are primarily based on GC/MS analysis of two selected samples [*18*], as well as comparison with GC/MS literature data on Wilsonville ITSL CDLs [*19–20*]. Moreover, the authors are aware of the probable presence of many other compound series with similar nominal mass values as the compound series tentatively identified in Fig. 3. Therefore, most chemical labels should be seen as an attempt to provide a "most probable identity" for the major compound series thought to be represented by a given series of mass peaks. As seen in Fig. 3*a,* the hydrotreater feed spectrum shows a relatively prominent ion series at *m/z* 168, 182, 196, 210, and 224, tentatively identified as acenaphthenes with various degrees of alkyl substitution and some contribution from tetrahydrophenanthrenes. The hydrotreater product (flashed bottoms) spectrum in Fig. 3*b* shows a major peak series at *m/z* 186, 200, 214, and 228 believed to represent hydroaromatic hydrocarbons (such as octahydrophenanthrenes or hexahydrofluorenes or both with possible contributions from tetrahydroacenaphthenes). Another major homologous ion series at *m/z* 212, 226, and 240 in the hydrotreater product spectrum is thought to represent decahydropyrenes (perhaps with contributions from decahydrofluoranthenes). Comparison of the hydrotreater feed and hydrotreater product spectra in Fig. 3 with the corresponding spectra in Fig. 4 confirms that higher molecular weight components condensed out in the reaction tube at ambient temperature inlet conditions. Nevertheless, close examination of the corresponding spectra obtained under the two different conditions (Figs. 3*a,b* and 4*a,b*) show that the same major homologous ion series discussed in previous paragraphs are also present in the preheated spectra and show the same characteristic differences between hydrotreater feed and hydrotreater product spectra.

Since the extreme complexity of the spectra in Fig. *4* makes it difficult to draw any conclusions about the chemical composition or the effect of hydrotreatment, advanced computerized data analysis techniques need to be called upon. Unfortunately, sophisticated data analysis reduction methods such as factor analysis cannot be readily applied to mass spectral patterns containing many hundreds of mass variables due to the size limitations of the files that can be handled by existing multivariate statistical analysis program packages. Thus, some of the more frequently used procedures for reducing the number of mass variables are: (1) random selection of variables, (2) selection of variables on the basis of univariate statistical analysis criteria [*11*], and (3) reduction of the mass range scanned. None of these procedures are entirely satisfactory from a multivariate analysis point of view. However the third procedure can be implemented conveniently by reducing the inlet temperature as shown in Fig. 3, while simultaneously reducing the rate of ion source contamination and also simplifying the data interpretation.

In order to enable a more systematic comparison between the spectra in the two data sets (ambient inlet temperature and preheated inlet), both data sets were subjected to factor analysis separately using the SIGMA Program [*12*] that was developed to get around the limitations of most standard multivariate program packages, which allows us to handle maximally 100 to 200 variables. In contrast, SIGMA allows up to 1000 variables when limiting the number of spectra to 30. Factor analysis results are given in Tables 2 and 3. The score plots of the first two factors are shown in Figs. 5*a* and 6*a* for ambient inlet and preheated inlet data, respectively. Two distinct clusters were obtained for both data sets, namely, a hydrotreater feed cluster (Sample Categories 1 to 10) in the Factor I+ and a hydrotreater product cluster (Sample Categories 11 to 20) on Factor I−. In order to maintain consistency between the two data sets, the numbering system is kept the same for the preheated inlet data.

Visual examination of the factor score plots in Figs. 5*a* and 6*a* shows that Factor I of the ambient temperature data set (62% total variance) and Factor I of the preheated inlet data set (64% total variance) represent the major "hydrotreatment effect" trend in both data sets.

TABLE 2—*Factor analysis results of ambient inlet temperature data.*

Factor	Eigenvalue	Total Variance, %	Cumulative Variance, %
1	59.11	62.35	62.35
2	10.47	11.05	73.40
3	7.07	7.46	80.86
4	2.44	2.57	83.43
5	2.29	2.24	85.85
6	1.54	1.62	87.47
7	1.22	1.28	88.76
8	1.13	1.19	89.94
9	1.07	1.13	91.07

TABLE 3—*Factor analysis results of preheated inlet (200°C) data.*

Factor	Eigenvalue	Total Variance, %	Cumulative Variance, %
1	96.55	64.37	64.37
2	24.09	16.06	80.43
3	10.22	6.81	87.24
4	6.84	4.56	91.80
5	4.38	2.92	94.72
6	4.22	2.81	97.53
7	3.70	2.47	100.00

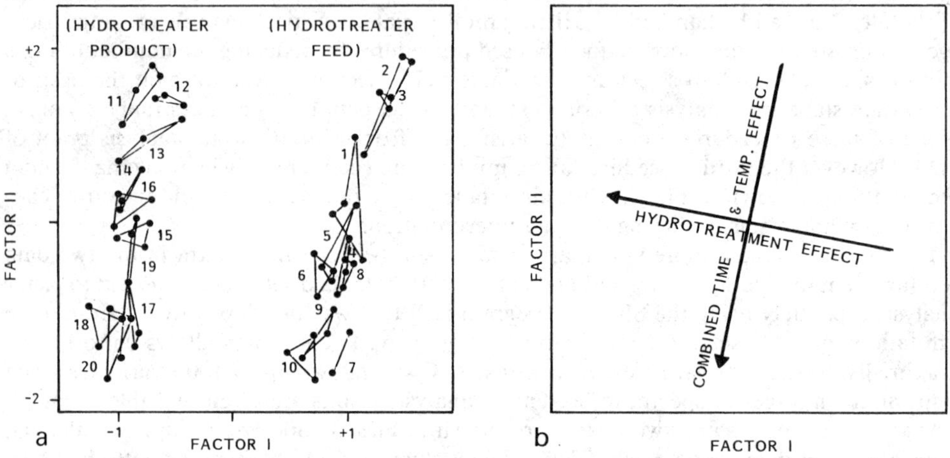

FIG. 5—*Factor score plot of* (a) *hydrotreater feed (Categories 1 to 10) and hydrotreater product samples (Categories 11 to 20) under ambient inlet temperatures, as described by Factor I and Factor II and* (b) *major trends found in Factor I and Factor II space.*

Factor II in the ambient inlet temperature data set (11% total variance) reveals a second trend in the data set, namely, an inferred "time + temperature effect." The two variables, process time and reactor temperature, are positively correlated due to the experimental design, as discussed in more detail elsewhere [*9*]. Therefore, it is not possible to distinguish and study the individual effects of these two parameters.

Note that the second trend (time + temperature effect) was not clearly observed in the preheated inlet data set. This is thought to be due to the limited number of samples analyzed or the large number of variables used or both. The trend or trends found in both data sets are shown in Figs. 5*b* and 6*b*. Only Factors I and II of the ambient inlet data and Factor I of the preheated inlet data revealed obvious chemical trends; the other factors did not show any interpretable trends or components in spite of various optimization attempts using nonsupervised factor rotation methods such as the Variance Diagram technique [*13–17*].

In order to translate the clustering effects or trends or both seen in Figs. 5*a* and 6*a* into mass spectrometric variables, so-called "factor spectra" were calculated according to a procedure described by Windig et al. [*13*] (Figs. 7, 8, and 9). The positive components of the Factor I spectrum at ambient inlet temperature (Fig. 7*a*) show the characteristic ion series

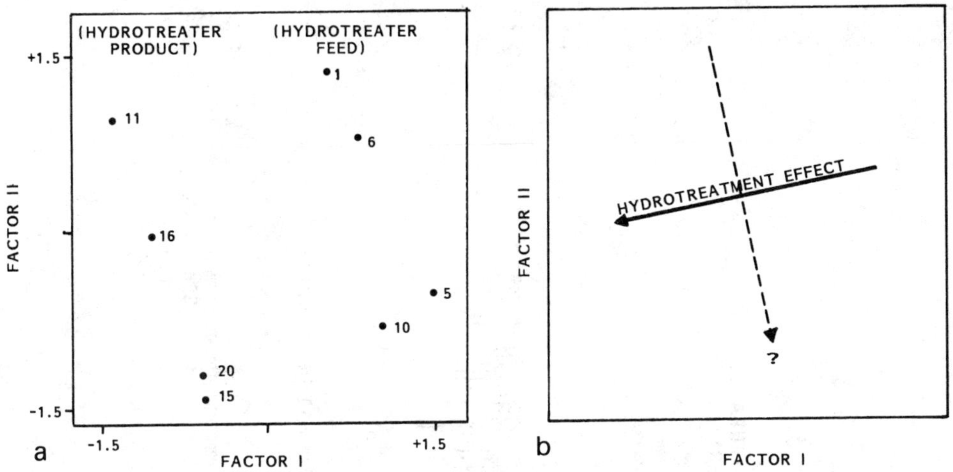

FIG. 6—*Factor score plot of* (a) *hydrotreater feed (Categories 1 to 10) and hydrotreater product (Categories 11 to 20) samples under preheated inlet conditions described by Factor I and Factor II and* (b) *major trend found in the Factor I and Factor II space.*

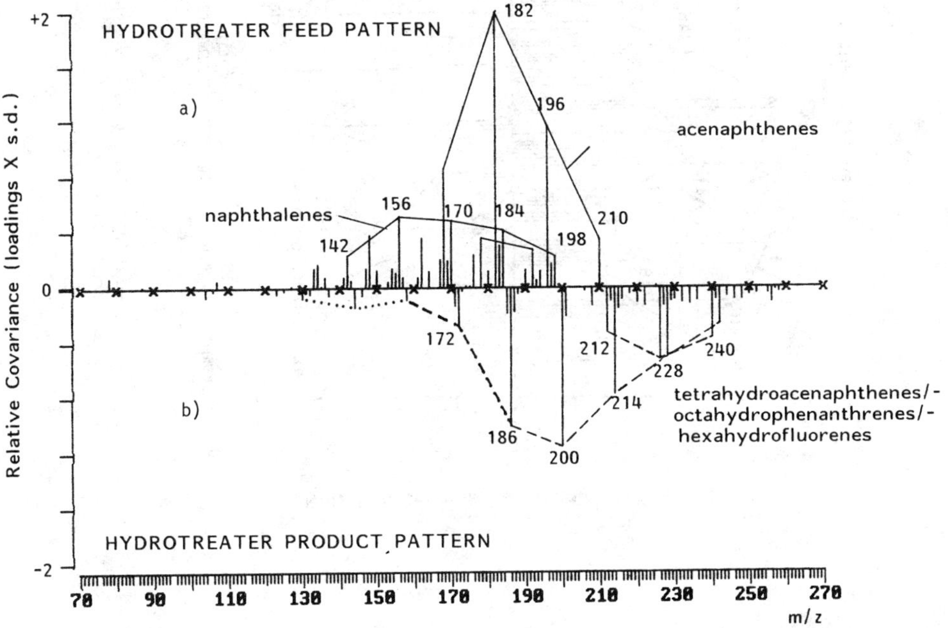

FIG. 7—*Factor spectra corresponding to the hydrotreatment axis (Fig.* 5b) *observed under ambient inlet conditions.*

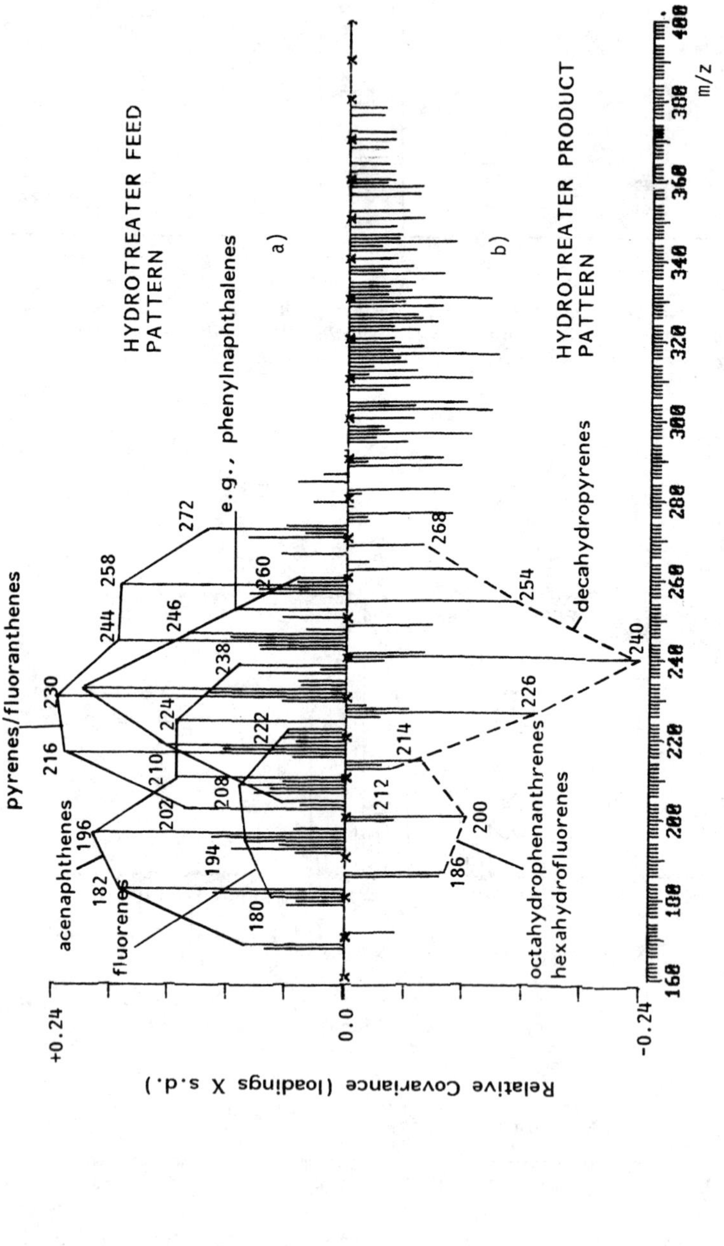

FIG. 8—*Factor spectra corresponding to the hydrotreatment axis (Fig. 6b) observed under preheated inlet conditions.*

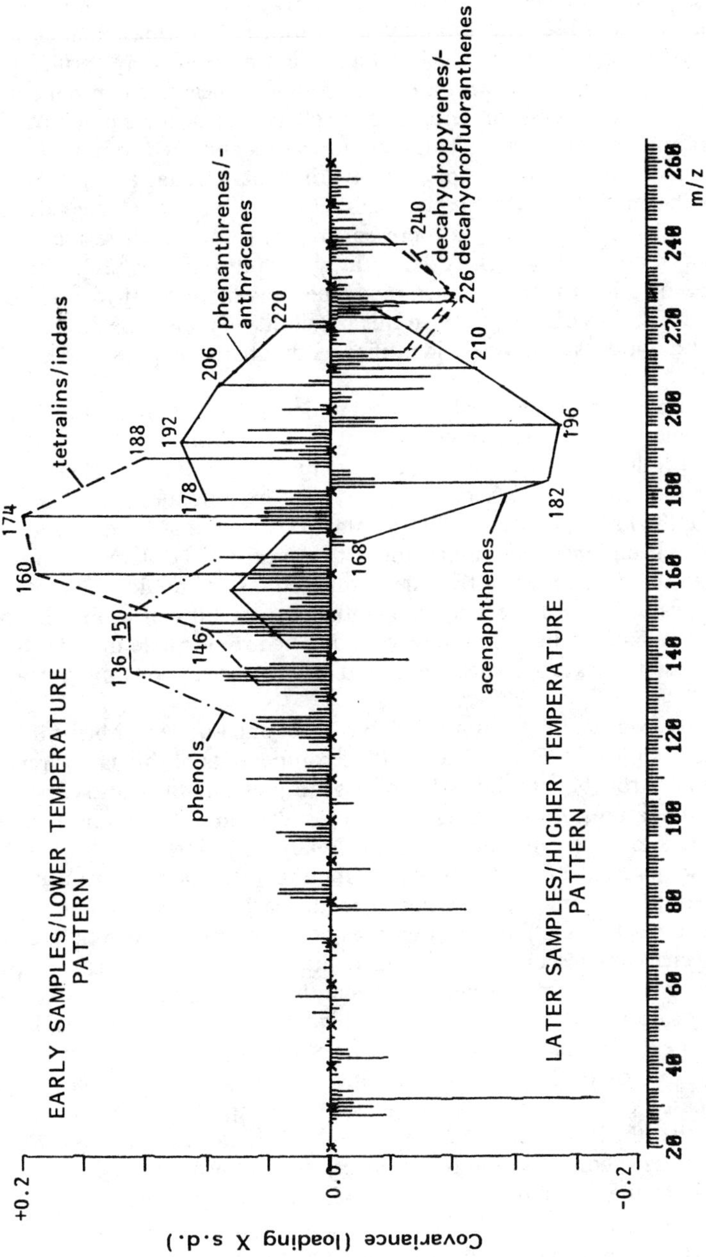

FIG. 9—*Factor spectra corresponding to the combined time + temperature trend observed under ambient inlet temperature conditions.*

representing the hydrotreater feed samples. As expected, these ion series correspond primarily to aromatic hydrocarbon compounds. For instance, the ion series at m/z 168, 182, 196, 210, and 224 can be tentatively identified to represent acenaphthenes or tetrahydrophenanthrenes or both, whereas other aromatic series at m/z 142, 156, 170, 184, and 198 and at m/z 134, 148, 162, 176, and 190 may be identified as alkylnaphthalenes and alkylbenzenes, respectively. As mentioned before, these identifications were verified by GC/MS analysis as well as by comparison with literature data. The negative component of Factor I, as shown in Fig. 7*b*, represents ion series that are characteristic for the hydrotreater product (flashed bottoms). Again, as expected, these ion series appeared to represent primarily hydroaromatic compounds, for example, octahydrophenanthrenes at m/z 186, 200, 214, 228, and 242 (with contributions from other possible compounds, as discussed before). The ion series at m/z 130, 144, etc., might represent fragment ions of these higher hydroaromatic compounds, due to retro-Diels-Alder pathway decompositions [*21*]. Alternatively, a small percentage of dihydronaphthalenes might be present in the hydrotreater product fraction, although GC/MS did not confirm this possibility. Finally, the major ion series at m/z 212, 226, and 240 was tentatively identified as decahydropyrenes or decahydrofluoranthenes or both.

The positive components of Factor I obtained under preheated inlet conditions (Fig. 8*a*) represent ion species that are relatively abundant in the hydrotreater feed as compared to the hydrotreater product, whereas the negative components of Factor I (Fig. 8*b*) represent the chemical compounds that are relatively high in the hydrotreater product.

Comparison of the Factor I spectra of both data sets (Figs. 7 and 8) reveals a high degree of qualitative correspondence in combination with the expected differences in relative ion intensities due to the differences in the inlet temperatures. It should be pointed out that the small ion series in Fig. 7, for example, naphthalenes, are absent in Fig. 8, since in the preheated inlet data a different selection of variables had to be made in order to allow for the presence of higher mass range variables found to be important by multivariate analysis criteria.

The numerically extracted spectrum of Factor II obtained under ambient inlet temperature conditions is shown in Fig. 9. The positive components of this factor represent ion series that correlate strongly with the early process samples and, thus, also correspond with lower reactor temperatures. As seen from the factor spectrum in Fig. 9, the average molecular weight appears to have shifted upward. The relative ion intensities of low molecular weight compounds present in relatively small amounts (for example, benzenes, phenols, naphthalenes, tetralins, or indanes) are higher in the early samples and at lower reactor temperatures, whereas the relative ion intensities of higher molecular weight components (for example, acenaphthenes or tetrahydrophenanthrenes or both, decahydropyrenes or decahydrofluoranthenes or both, octahydrophenanthrenes and hexahydrofluorenes have increased in later samples which corresponds to higher reactor temperatures. It seems logical to attempt to explain these results as a result of catalyst aging. The unavoidable loss of catalyst efficiency would be expected to result in loss of bond cleavage activity, thereby causing an increased overall molecular weight of the compounds formed. Alternatively, one might argue that the loss of efficiency of capping newly formed radicals during catalyst aging leads to increased condensation reactions of active radical species.

As just discussed, the ambient inlet data set shows the time and temperature trend, which is represented by Factor II (Fig. 5*a*). Unfortunately, this trend is not obtained by the preheated inlet data set, which might be due to limited number of samples analyzed. Therefore at this time, the chemical significance of the correlated variance represented by Factor II in Fig. 6a is unclear.

Conclusion

The results show the applicability of computer-assisted low-voltage MS techniques in the process monitoring of coal-derived liquids. Two major trends were found in the data set under ambient inlet temperature: (1) a hydrotreatment effect that was characterized by a shift from aromatic to hydroaromatic hydrocarbons, and (2) a time + temperature effect where the average molecular weight appeared to shift to higher molecular weight components as the catalyst aged and the temperature of the reactor increased.

The main hydrotreatment effect was also observed clearly in the preheated inlet data whereas the time + temperature effect was missed, apparently due to the limited number of samples analyzed.

The results also show that, although preheating the inlet is an effective way of extending the spectra into higher mass ranges, the structural assignments of the higher mass peaks become more and more difficult. Moreover, problems are encountered in handling wide mass range data with existing, standard statistical packages. A data set obtained under ambient inlet conditions shows essentially the same characteristic component types observed under preheated inlet conditions with the exception of the higher boiling components. Advantages of the ambient inlet conditions over preheated inlet condition are speed, ease of data handling and reduced ion source contamination, thus resulting in long-term reproducibility of the spectra.

Acknowledgments

This work was supported by the Commonwealth of Kentucky, Kentucky Energy Cabinet, and U.S. Department of Energy Contract No. DEFC22-85PC80009 as part of the Consortium for Fossil Fuel Liquefaction Science. The authors also wish to acknowledge the invaluable cooperation of the staff of the Wilsonville Alabama pilot plant.

References

[*1*] Taghizadeh, K., Hardy, R. H., Davis, B. H., and Meuzelaar, H. L. C., "Comparison of Low Voltage and Field Ionization Mass Spectra of Coal-Derived Liquids from the Wilsonville Pilot Plant," submitted to *Analytical Chemistry.*

[*2*] St. John, G. A., Buttrill, S. E., Jr., and Anbar, M. in *Organic Chemistry of Coal,* Larsen, J. W., Ed., American Chemical Society Symposium Series, Washington, DC, 1978, pp. 223–239.

[*3*] Anbar, M. and St. John, G. A., *Fuel,* Vol. 57, 1978, pp. 105–110.

[*4*] Yoshida, T., Maekawa, Y., and Shimada, T., *Analytical Chemistry,* Vol. 54, 1982, pp. 967–971.

[*5*] Boduszynski, M. M., Hurtubise, R. J., Allen, T. W., and Silver, H. F., *Analytical Chemistry,* Vol. 55, 1983, pp. 225–231.

[*6*] Boduszynski, M. M., Hurtubise, R. J., and Allen, T. W., *Analytical Chemistry,* Vol. 55, 1983, pp. 232–241.

[*7*] Aczel, T., Williams, R. B., Pancviov, R. J., and Karchmer, J. H., final report to U.S. Department of Energy, MERC-8007-1 (Part 1), Sept. 1976.

[*8*] Jakab, E., Windig, W., and Meuzelaar, H. L. C., *Energy and Fuels,* Vol. 1, 1987, pp. 161–167.

[*9*] Taghizadeh, K., Hardy, R. H., Keogh, R., Goodman, J., Davis, B. H., and Meuzelaar, H. L. C., American Chemical Society preprints, Division of Fuel Chemistry, Vol. 32, No. 3, 1987, pp. 332–341.

[*10*] Moniz, M. J., Pillai, R. S., Lee, J. M., Lamb, C. W., and Johnson, T. W. in *Proceedings,* Ninth Annual EPRI Contractors Conference on Coal Liquefaction, EPRI AP-3825-SR Special Report, Electrical Power Research Institute, March 1985, Chapter 20.

[*11*] Meuzelaar, H. L. C., Haverkamp, J., and Hileman, F. D., *Pyrolysis Mass Spectrometry of Recent and Fossil Biomaterials,* Elsevier Scientific Publishing Company, Amsterdam, 1982.

[*12*] Windig, W. and Meuzelaar, H. L. C. in *Computer-Enhanced Analytical Spectroscopy,* H. L. C. Meuzelaar and T. L. Isenhour, Eds., Plenum Press, New York, 1987, pp. 67–102.

[13] Windig, W., Haverkamp, J., and Kistemaker, P. G., *Analytical Chemistry,* Vol. 55, 1983, pp. 81–88.
[14] Windig, W., Kistemaker, P. G., and Haverkamp, J., *Journal of Analytical and Applied Pyrolysis,* 1981, Vol. 3, pp. 199–212.
[15] Windig, W., de Hoog, G. S., Haverkamp, J., *Journal of Analytical and Applied Pyrolysis,* 1981, Vol. 3, pp. 213–220.
[16] Windig, W. and Meuzelaar, H. L. C., *Analytical Chemistry,* 1988, Vol. 56, pp. 2297–2303.
[17] Metcalf, G. S., Windig, W., Hill, G. R., and Meuzelaar, H. L. C., *International Journal of Coal Geology,* Vol. 7, 1987, pp. 245–268.
[18] Unpublished results.
[19] Wozniak, T. J. and Hites, R. A., *Analytical Chemistry,* Vol. 57, 1985, pp. 1320–1327.
[20] Winschel, R. A. and Burke, F. P., *Proceedings,* Eighth Annual EPRI Contractors Conference on Coal Liquefaction, EPRI AP-3366-SR. Special Reports, Electrical Power Research Institute, Feb. 1984, Chapter 16.
[21] Wozniak, T. J. and Hites, R. A., *Analytical Chemistry,* Vol. 57, 1985, pp. 1314–1319.

Thomas Aczel,[1] S. G. Colgrove,[2] and Lan Le[3]

An Expert Program for the Interpretation of Gas Chromatography/Mass Spectrometry Data on Fossil Fuel Distillates

REFERENCE: Aczel, T., Colgrove, S. G., Le, L., **"An Expert Program for the Interpretation of Gas Chromatography/Mass Spectrometry Data on Fossil Fuel Distillates,"** *Novel Techniques in Fossil Fuel Mass Spectrometry, ASTM STP 1019,* T. R. Ashe and K. V. Wood, Eds., American Society for Testing and Materials, Philadelphia, 1989, pp. 159–171.

ABSTRACT: An expert system has been developed for the interpretation of the gas chromatography/mass spectrometry (GC/MS) spectra of fossil fuel distillates. The approach extends the scope of commercially available data systems by using interpretation rules generally carried out by a skilled professional expert. The rules and procedures incorporated in the program deconvolute unresolved peaks, set acceptability limits for the identifications provided by commercial data systems, and identify components that the commercial data systems fail to identify, or identify incorrectly, at least as to compound class and carbon number.

Need for such an approach stems from the inability of commercial data systems to deal adequately with complex hydrocarbon mixtures. The spectral matching algorithms used by these systems do not consider spectra/structure correlations and fail when GC/MS peaks are unresolved or pure compound spectra are not available for comparison or both. These conditions occur in the majority of cases with boiling range above 149°C (300°F). The expert program discussed in the paper obviates these difficulties. It results in 90 to 95% correct identifications, 99% plus on a concentration basis, and it saves 20 to 30 h of skilled time per sample.

KEY WORDS: fossil fuels, mass spectrometry, gas chromatography, fossil fuel distillates

Gas chromatography/mass spectrometry (GC/MS) is one of the most useful analytical tools for the characterization of fossil fuel distillates. The method combines the separation capabilities of the gas chromatograph with the identification capabilities of a mass spectrometer and has been widely used for the last several years. The most efficient use of this instrumentation relies on computerized data systems, usually furnished by the instrument manufacturer. For component identification, these rely principally on comparing the spectra of sample components with those in a library of pure compound spectra. In spite of the rather extensive computer capabilities of most commercial data systems, interpretation of the GC/MS spectra is still very labor intensive, in particular for non-routine samples containing hundreds of components, such as fossil fuel distillates. Spectra have to be examined individually by a skilled analytical chemist and quantification is very cumbersome.

This situation is ideally suited for the development of an "expert" system, that takes advantage of the capabilities of the commercial software, and that, in addition, automates

[1] Exxon Research and Engineering Company, Annandale, NJ 08881.
[2] Exxon Research and Development Laboratories, Baton Rouge, LA.
[3] University of Houston, Chemistry Dept., University Park, Houston, TX.

most of the decision steps involved in human interpretation. Such a program has been developed at Exxon Research and is showing a very high degree of reliability.

Summary

An expert system has been developed for the interpretation of the GC/MS spectra of fossil fuel distillates. The approach uses commercially available identification capabilities, but it extends those by defining limits of acceptability for the library identification yielded by those systems. These limits are based on automatic examination of the spectra, on retention time considerations and on the quality of the library matching itself. Components with "unacceptable" library identifications are still identified at least in terms of chemical type and carbon number. The system allows the deconvolution of GC peaks containing multiple components (the majority) and includes quantification and sorting capabilities. Initial applications show 90 to 95% correct identifications in samples with up to 800 components.

Discussion

Need for an Expert System

The GC/MS spectra of fossil fuel distillates are extremely complex. Samples in the boiling range of interest, 38 to 260°C (100 to 500°F), contain up to 800 to 1000+ components, many of them isomers with very similar spectral features. "Manual" identification of this many components, even with the aid of computerized library searching systems, requires a highly skilled professional and is extremely time consuming. In depth interpretation requires hours and even days, and quantification is even more lengthy. This labor-intensive effort acts in effect as a hurdle and hinders the use of GC/MS for the analysis of complex fossil fuel distillates, resulting either in fewer samples being analyzed or in partial interpretation in which only the major components (or those *a priori* defined as the most important) are considered. Decrease of the skilled effort involved would obviously greatly increase the usefulness of the GC/MS technique.

There are additional compelling reasons to extend the capability of commercially available data systems for component identification.

1. Identification is based on the statistical matching of sample and library spectra, and does not involve "reasoned" consideration of spectra/structure correlations. Erroneous identifications are frequent.
2. There are no pure compound spectra available for most hydrocarbon components. No acceptable identification is available in these cases from the data system.
3. Spectra of many isomers are very similar; replicate identifications are also frequent. This is particularly true in cases where the library contains the spectra of only one of the many possible isomers.
4. Many components are unresolved, compounding the difficulties of identifications based exclusively on spectral matching, even when utilizing the deconvolution algorithms available in commercial data systems.

Given the preceding restraints, it is evident why human identification is so time consuming, even with the aid of the data systems. It has to be based on a large number of experience based, interconnected decisions, that include detailed examination of the mass spectra, considering the relative order of elution, and a very critical evaluation of the library response. On the other hand, these interpretation requirements are well suited to the capabilities of an expert program, that indeed can "mimic" the approach and decision-making abilities of the human expert. Such a program, through its capability to routinely invoke isotope correction and mixture analysis algorithms, can, in fact, even improve on human interpretation and, in any case, be more consistent.

Expert Systems

An expert system is a computer program that simulates, within certain limitations, the specialized knowledge of human experts to solve a particular type of problem. Expert systems consist of two major components, as shown in Fig. 1. The knowledge base contains a set of rules (and possibly facts) in the specific area of expertise. The inference engine contains the inference strategies and controls the reasoning process. Problems are presented to the system by the user in terms of data that are known about a particular situation. Reasoning is achieved by applying the facts and rules that are stored in the knowledge base to the data acquired from the user.

From years of experience, experts have learned many tricks and shortcuts that often are never written down. These are known as "heuristics." Heuristics allow experts to make good educated guesses that often lead quickly to a solution. Although experts have a broad understanding of the problems in their areas of expertise, they often cannot easily explain

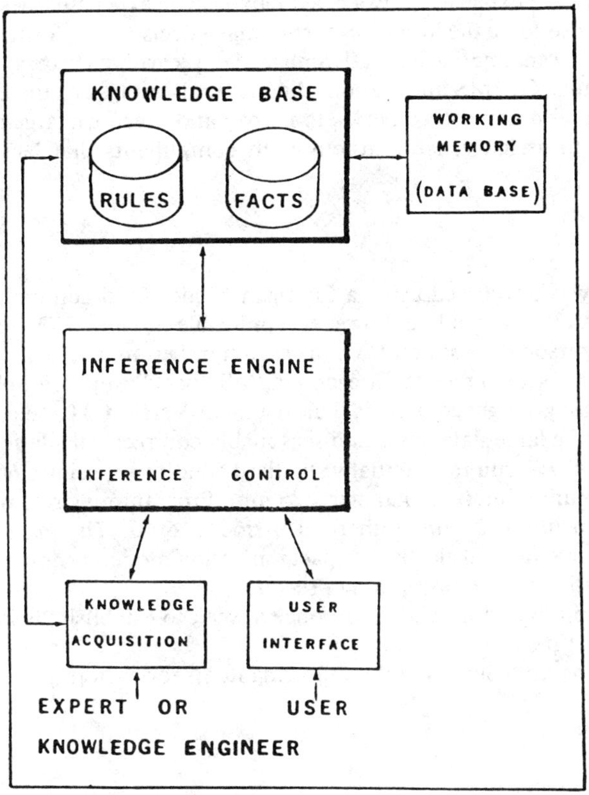

FIG. 1—*Schematic of expert systems.*

how they solve problems in these areas. They usually do not structure and arrange their knowledge so that it can be described in a systematic way.

Over the past few years, the new discipline of "knowledge engineering" has appeared. After carefully defining the problem to be solved and specifying the requirements to be met, the knowledge engineer elicits and collects the heuristics of experts in a field and structures that expertise in a decision-making package, an expert system.

The knowledge engineer and the expert work together continuously so that existing information may be incorporated and new information may be added to the knowledge base. The knowledge engineer becomes intimately familiar with the expert system and its subject matter.

When a number of rules have been established, the prototype expert system is tested against cases with known outcomes. During this phase of the expert system development, rules and facts are refined. Finally, the expert system is tested against cases with unknown outcomes and the results are checked by the expert to validate the expert system.

Expert systems tend to be best at solving problems involving deductive reasoning. This situation arises, for example, when the rules are expressible in terms of IF antecedents and THEN or ELSE consequents. Although expert systems can be written using almost any computer language, LISP (List Processing) is commonly considered to be one of the best languages for any computer work. Other languages such as BASIC, FORTRAN, and PASCAL can be used to construct rule interpreters and to create a rule base.

Expert systems have several advantages with respect to human experts. They are more consistent, faster, less expensive, always available (their expertise does not get lost), easy to change. They also force the human expert to make decisions or formulate rules in a very well defined, logical manner; a great advantage and a learning process itself.

For the particular GC/MS application, the main advantages of the expert program are its ability to deal with unresolved peaks; to invoke and execute a large number of analysis algorithms; and to assign general structures to components that lack specific structure identifications.

Experimental

The program was developed using a Finnigan Model 4025 automated GC/MS system and an IBM PC/AT. It should be, however, applicable to any GC/MS system that is able to transmit to a personal computer (PC) the scan number, mass spectrum, total ionization current, and library search results for each GC/MS "peak" in the sample.

The programming language used is Turbo Pascal Version 3.0, selected because of its capability to handle large data bases and reasonably complex calculations.

The GC/MS/DS was run in essentially standard conditions, using a 60-m J and W fused silica capillary column, internal diameter 0.25 mm. Programming rate was 4 cpm, from 50 to 350°C, with an initial 2 min isothermal period at 50°C. The mass spectrometer was operated in the conventional electron impact ionization mode, scanning a mass range from m/z 29 to 500 with a rate of one scan per second.

A normal paraffin blend C_6–C_{40}, is run once a week to establish "historical" data for *n*-paraffin retention times.

Input to the program consists of the following items for each peak:

1. Run number.
2. Scan number.
3. Total ionization (RIC).
4. Mass spectrum, m/e, and intensity values.

5. Results of the library search, including name, formula, "fit," and "purity" for the nine possible components found by the search routines. Library searches might be restricted to specific classes such as saturate and aromatic hydrocarbons.

A schematic illustration of the input and the overall data flow is given in Fig. 2.

The programs include a "filter" routine to eliminate unnecessary header and search information. Mass spectra are isotope corrected for contributions to M + 1, M + 2, and M + 3 ions from carbon, hydrogen, oxygen, sulfur, and nitrogen isotopes. This step increases the accuracy of the subsequent spectral work-up and interpretation.

Calibration matrices for the various type analyses are stored on disk as well as tables relating the boiling points and expected retention types of saturate and aromatic hydrocarbons to those of *n*-paraffins.

Main Principles of Expert Logic

The key feature of the expert program is to treat each GC/MS peak as a simple mixture, and to deconvolute it using conventional MS type analyses. This approach quantitates the type assignment process used by an expert who relies on the examination of typical fragmentation patterns. The type analysis is followed by the definition of the range of acceptable carbon numbers for each of the major types, the search for molecular ions, and the comparison of the limits so established with the library responses. This process, and the rules used at present, is discussed in some detail in the following paragraphs. A summary is given in Table 1.

Preliminary Identification of N-*Paraffins*—Normal paraffin carbon numbers are used to select the appropriate type analyses in a given retention time range. This requires the approximate identification of their retention times. This is done using values calculated

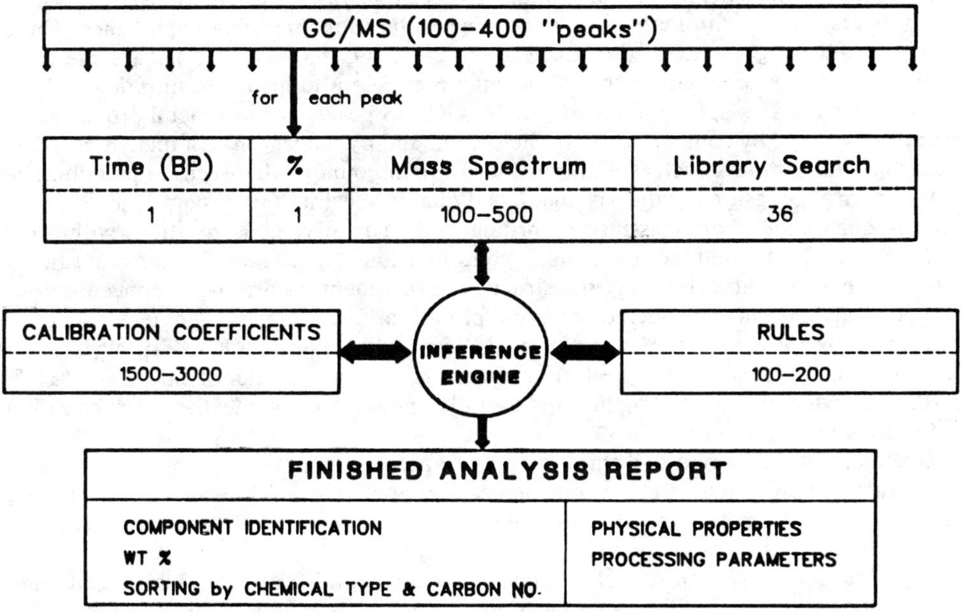

FIG. 2—*Overview of specific expert system for gas chromatography-mass spectrometry.*

TABLE 1—*Sequence of steps in expert logic.*

1. Tentative identification of *n*-paraffins; selection of appropriate carbon number inverse for type analysis.
2. Type analysis; selection of candidates for specific structure assignment.
3. Definitive identification of *n*-paraffins, if present in sample.
4. Assignment of acceptable carbon number ranges for each type.
5. Search for molecular ions.
6. Component identifications:
   a. assignment of specific structure, if warranted; and
   b. assignment of general structure (type, carbon number).
7. Quantitation.
8. Sorting according to type, carbon number.
9. Report generation.

from previous (historical) runs of a standard; retention time windows of optional width are assigned to each prospective *n*-paraffin.

Type Analyses—Type analyses are carried out on the spectra of each GC/MS peak, using any of several literature or proprietary methods. For light distillates that are of prevalently saturate character ASTM Method for Hydrocarbon Types Analysis of Gas-Oil Saturate Fractions by High Ionizing Voltage Mass Spectrometry (D 2786-86) is used; ASTM Test Method for Hydrocarbon Types in Low Olefinic Gasoline by Mass Spectrometry (D 2789-86) for samples suspected to contain naphthalenes; and others are being continuously added. To date, ASTM D 2786-86 has been the most frequently used one; it yields relative percentages of paraffins, one to six ring naphthenes, and monoaromatics, namely, benzenes, indans plus tetralins, and indenes plus dinaphthenobenzenes. Matrix, or rather inverse selection, is based on the carbon number of the putative adjacent *n*-paraffin, on the high retention time side. Spectra are isotope corrected before the calculations; this precaution was required to eliminate erroneously high values for three-ring naphthenes. These are calculated using characteristic peak heights at *m/e* 149, 163, 177, etc., that are interfered by the ^{13}C isotopes of alkylbenzenes if the latter are present in high concentrations.

Use of type analyses, for the much simpler GC/MS peaks, has in general proved satisfactory, as proven by comparison with the spectra and by the fact that in many cases they yield high concentration values, 98% plus, for the compound(s) that seem to predominate in the spectra. Occasional failure is due to bad quality spectra of trace components.

Components found are classified according to their relative concentrations within the GC/MS peak. At present, components in concentrations higher than 50% are considered prospective candidates for specific structure assignment; those with concentrations between 5 and 50% are considered only for type and carbon number assignment; and those with concentrations below 5% are neglected. Because the average GC/MS peak is about 0.2 to 0.4% of a typical sample, neglecting concentrations of the order of 0.01 to 0.02% (5% of 0.2 to 0.4) does not adversely affect the overall accuracy. At any rate, these concentration limits are optional input values.

Treatment of the second and third largest components as potential candidates for specific structure assignment is being considered. This approach would involve correction of the spectra by subtracting the contributions of the identified components. Results of the type analysis are also used in the subsequent quantitation step.

Identification of N-*Paraffins*—Normal paraffins are identified using both retention times and spectral features. Positive identification requires satisfaction of all of the following criteria.

1. Retention time must lie within the time window established using historical values.
2. Type analysis must show at least 80% paraffin content.
3. Major fragment peaks should follow the characteristic pattern for *n*-paraffins, that is, m/e 57 $<$ four times m/e 71; m/e 71 $>$ m/e 85 $>$ m/e 99 $>$ m/e 113 $>$ m/e 127.
4. Retention times of identified *n*-paraffins must have the same time ratios to other *n*-paraffins as the ratios found for the corresponding components in a standard *n*-paraffin blend.

Failure to identify *n*-paraffins in the sample has no negative consequence on the identification of the other components; it might, however, trigger checks on instrument performance (temperature programming rate, column degradation, etc.). This is planned to be an interactive process, involving information exchange between operator and program on both sample history (Are *n*-paraffins likely to be present in this product?) and instrument parameters (If *n*-paraffins are likely to be present, and if program could not correctly find them, are then temperature programming, column conditions, etc. within specifications?).

Assignment of Hydrocarbon Carbon Numbers—Major types determined by the type analysis are assigned a range of acceptable carbon numbers. Assignment is based on retention time with respect to the adjacent normal paraffins and on the relative boiling range of a particular chemical class with respect to normal paraffins. For example, isoparaffins are assigned acceptable carbon numbers that range from *n*-2 to *n* if their retention time is between a normal paraffin of carbon number *n*-1 and *n*; alkyl benzenes in the same retention time region are assigned carbon numbers from *n*-3 to *n*-1. A table with these assignments is stored in the program; one of the three acceptable carbon numbers is designed as the "most probable" one.

Subsequent searches for molecular ions are limited to the types given by the type analysis and to the acceptable carbon number range as defined in this section.

Search for Molecular Ions—Molecular ions are searched for in a manner similar to what an expert would do. The search is limited to m/e values defined by the type (Z number) and the carbon number(s) as determined in the previous steps. For example, molecular ion values for two-ring naphthenes, Z number $= -2$, with retention time between C_{12} and C_{13} *n*-paraffin are

$$C_{11}H_{20}{:}152$$
$$C_{12}H_{22}{:}166$$
$$C_{13}H_{24}{:}180$$

The simple arithmetic involved to convert Z numbers and carbon numbers to masses is of course executed by the programs.

For monoaromatics, as determined by ASTM D 2786-86, the program calculates three masses for each of the types included, that is alkylbenzenes, indans or tetralins, indenes or dinaphthenobenzenes, plus for naphthalenes that are assumed to be potentially present in peaks with high aromatic content.

In order to be accepted as bona fide molecular ions, the "acceptable" masses, M, have to satisfy the following criteria.

1. Intensity should be isotope free, to discriminate against ^{13}C isotopes of large fragment peaks. The isotope correction procedure is carried out on each spectrum, as discussed previously.

2. Intensity of M should be higher than that of M + 1. This criterion eliminates most fragment ions.
3. Intensity of M should be more than 10% of highest intensity molecular ion within the given GC/MS peak. This criterion eliminates molecular ions that are present in trace concentration. The 10% threshold is optional, and it can be changed as required.
4. For naphthenes only, the intensity distribution of the molecular ions, if more than one is present, should be gaussian or monotonic. Two or three molecular ions, M_1, M_2, or M_3 in the same Z series are accepted only if the following relationships are true:
   (*a*) $M_1 + 14 = M_2$ (the two molecular ions have carbon numbers differing by only one),
   (*b*) $M_1 < M_2 > M_3$,
   (*c*) $M_3 < M_2 > M_1$,
   (*d*) $M_1 > M_2 > M_3$, and
   (*e*) $M_1 < M_2 < M_3$.

These rules eliminate pseudo-molecular ions due to H rearrangement, that occur frequently in naphthenes (and olefins) at masses corresponding to M-14, M-28, etc. In effect, the program accepts only gaussian or uniformly changing distributions, as one expects in distillation fractions. A GC/MS peak is obviously a model for a distillation fraction.

Structure Assignment—Structure assignments are made with the aid of the commercial data systems. Acceptance of library identifications is, however, strictly limited by the criteria or boundaries defined in the previous steps. Only those structures are accepted that are of the chemical class determined by the type analysis and that have a carbon number (formula) within the limits defined by retention time and type.

In addition, the programs also use criteria based on the quality of the library responses. These are optional inputs that set thresholds for the "fit" and "purity" parameters that the commercial programs use to evaluate the "quality" of the spectral matches. At present, the program requires a fit value of at least $^{900}/_{1000}$ for consideration of a library identification. In the case of more than one identification with fit values above $^{900}/_{1000}$, the program selects the one with the highest fit value. No specific structures are assigned if there is no structure associated with the same or higher fit value than that specified in the threshold.

TABLE 2—*Summary of rules for assignment of specific structures.*

Found		
Molecular Ion(s)	Acceptable Structure	Decision
No	No	Assign type, "most probable" carbon number for each type
No	Yes	Assign structure, corresponding carbon number, type
Yes	No	Assign type, carbon number to molecular ions found
Yes	Yes	Match: assign structure; select best "fit" if more than one match No match: assign type, carbon number of molecular ion found; declare conflict

The cases and rules considered in the assignment of the specific structures are summarized in Table 2. As shown, each component is assigned at least a chemical formula and a carbon number; many are assigned specific structures.

Quantitation and Sorting—All components are quantitated, using the relative abundance of the GC/MS peaks, defined as the ratio of their reconstructed ion currents (RIC) to the sum of all the RIC values in a given sample.

TABLE 3—*Partial list of components in a kerosene.*

Retention Time	Mnemonic	Carbon No.	Boiling Point, °F	Percent by Weight	Identification
1053	2RNP	11	376.48	0.03	2-ring naphthene (assigned type)
1053	BENZ	11	376.48	0.12	alkylbenzene (assigned type)
1056	1RNP	11	377.00	0.05	1-ring naphthene (assigned type)
1056	2RNP	11	377.00	0.04	2-ring naphthene (assigned type)
1056	INDEN	9	377.00	0.03	indene (assigned type)
1056	INDAN	11	377.00	0.04	indan/tetralin (assigned type)
1058	BENZ	10	377.35	0.12	benzene, 1-methyl-4-(1-methylethyl)-
1058	INDAN	11	377.35	0.03	indan/tetralin (assigned type)
1058	BENZ	11	377.35	0.18	alkylbenzene (assigned type)
1061	1RNP	11	377.87	0.22	1-ring naphthene (assigned type)
1065	1RNP	11	378.56	0.09	1-ring naphthene (assigned type)
1067	1RNP	11	378.90	0.05	1-ring naphthene (assigned type)
1067	2RNP	11	378.90	0.02	2-ring naphthene (assigned type)
1067	INDEN	9	378.90	0.03	indene (assigned type)
1072	1RNP	11	379.76	0.02	1-ring naphthene (assigned type)
1072	BENZ	9	379.76	0.01	alkylbenzene (assigned type)
1072	BENZ	11	379.76	0.11	alkylbenzene (assigned type)
1074	IPAR	11	380.11	0.86	iso-paraffin (assigned type)
1076	NPAR	11	380.46	2.27	undecane
1081	IPAR	11	381.32	0.01	iso-paraffin (assigned type)
1081	1RNP	11	381.32	0.01	1-ring naphthene (assigned type)
1081	2RNP	11	381.32	0.11	2-ring naphthene (assigned type)
1081	BENZ	11	381.32	0.03	alkylbenzene (assigned type)
1085	BENZ	9	382.01	0.01	alkylbenzene (assigned type)
1085	BENZ	10	382.01	0.06	alkylbenzene (assigned type)
1085	BENZ	11	382.01	0.05	alkylbenzene (assigned type)
1087	BENZ	10	382.36	0.13	benzene, 1,4-diethyl-
1093	IPAR	11	383.39	0.34	iso-paraffin (assigned type)
1093	1RNP	11	383.39	0.04	1-ring naphthene (assigned type)
1093	2RNP	11	383.39	0.36	naphthalene, decahydro-2-methyl-
1093	BENZ	10	383.39	0.13	alkylbenzene (assigned type)
1096	BENZ	10	383.91	0.20	benzene, 1,2,3,4-tetramethyl-
1102	IPAR	11	384.90	0.07	iso-paraffin (assigned type)
1102	2RNP	10	384.90	0.04	2-ring naphthene (assigned type)
1115	IPAR	12	386.87	0.65	decane, 3,6-dimethyl-
1115	1RNP	12	386.87	0.15	1-ring naphthene (assigned type)
1115	2RNP	12	386.87	0.08	2-ring naphthene (assigned type)
1129	IPAR	12	388.98	0.35	iso-paraffin (assigned type)
1129	1RNP	12	388.98	0.13	1-ring naphthene (assigned type)
1129	2RNP	11	388.98	0.48	naphthalene, decahydro-2-methyl-
1129	BENZ	11	388.98	0.08	alkylbenzene (assigned type)
1138	IPAR	12	390.34	0.97	undecane, 2-methyl-
1145	BENZ	11	391.40	0.11	benzene, (1,1-dimethylethyl) methyl-

Deconvolution of each GC/MS peak is carried out using the following rules if required.

1. The concentration of each GC/MS peak is prorated amongst the compound types that it contains. Types below an optional threshold, 5% at present, are neglected, and their concentrations are prorated amongst the other types.
2. Concentration of each type is prorated amongst the molecular ions, if any, of that type. Equal sensitivity is assumed among molecular ions of the same type. In this context, monoaromatics determined by ASTM D 2786-86 are considered as one type.
3. In the absence of molecular ions, monoaromatics are distributed into the component

TABLE 4—*Partial example of sorting by type, carbon number, individual isomers.*

Components	GC/MS	GC/MS Norm to LR/MS
ISOPARAFFINS (con't)		
C13 (con't)		
Isomer L	0.01	0.01
Isomer M	0.01	0.01
Isomer N	0.01	0.01
C14		
Isomer A	trace	trace
1-RING NAPHTHENES	35.08	33.15
Cyclopentanes		
C9		
Butylcyclopentane	0.14	0.13
C10		
Pentylcyclopentane	0.57	0.54
Cyclohexanes		
C9		
1,3,5-trimethylcyclohexane	0.02	0.02
1,1,2-or 1,1,3-TrimethylCH	trace	trace
a Trimethylcyclohexane	0.02	0.02
a Trimethylcyclohexane	0.01	0.01
a Trimethylcyclohexane	0.02	0.02
1,2,4-Trimethylcyclohexane	0.05	0.05
a Trimethylcyclohexane	0.02	0.02
an Ethylmethylcyclohexane	0.06	0.06
an Ethylmethylcyclohexane	0.04	0.04
Isomer A	0.01	0.01
an Ethylmethylcyclohexane	0.07	0.07
an Ethylmethylcyclohexane	0.04	0.04
Propylcyclohexane	0.31	0.29
C10		
Diethylcyclohexane	0.31	0.29
Ethyldimethylcyclohexane	0.20	0.19
Methylisopropylcyclohexane	0.51	0.48
Methylpropylcyclohexane	1.67	1.57
Butylcyclohexane	0.73	0.69

benzenes, indans plus tetralins, and indenes plus dinaphthenobenzenes, using the relative intensities of their characteristic fragment peaks.

A partial example of the output of the identification and quantitation process is given in Table 3. Each line corresponds to a component; parameters given are:

1. Retention time.
2. Mnemonic (a four character word for each type, later used in sorting).
3. Carbon number.
4. Boiling point (calculated using *n*-paraffin calibration).
5. Percent by weight.
6. Specific or generic structure (name).

The advantage of this output is that it reduces the normal output of the manufacturers data system, two or three pages per component (often a mixture) to one line for each resolved component. A "book" of a thousand or more pages is thus reduced to 10 to 20 easily readable pages. Detailed output, including the spectra, type analysis result, library information, and, above all, the decision steps involved in the identification process, is also available as an option. It can be recalled on the PC screen by invoking the retention time.

Two types of sorting are available.

In one, components are sorted according to type, carbon number, and individual isomer. Each isomer is listed individually; it is assigned either a specific structure, or at least a boiling point (or retention index). In this way, the presence of even not specifically identified isomers can be compared from sample to sample (Table 4).

The second type of sorting involves the summing of isomers. The output is displayed as a compound type, carbon number matrix (Table 5). At this point, the sorting involves about 20 different types, including paraffins, cyclopentanes, cyclohexanes, one to four-ring naphthenes, and several olefinic and aromatic types.

Calculation of physical properties, such as elemental analysis, average carbon number,

TABLE 5—*Partial illustration of sorted output.*

	Percent by Weight					
Carbon No.	*n*-Paraffins	*I*-Paraffins	Cycloparaffins	Benzenes	Indans	Total
8	3.2	2.1	8.4	3.8	. . .	17.5
9	4.7	3.7	10.3	6.6	1.0	26.3
10	6.8	4.2	7.6	3.2	0.6	22.4
11	5.3	3.8	5.4	2.1	0.4	17.0
12	2.1	1.9	3.7	1.5	0.0	9.2
13	1.6	1.4	2.1	0.6	0.0	5.7
14	0.7	0.6	0.3	0.0	0.0	1.6
15	0.2	0.1	0.0	0.0	0.0	0.3
Total	24.6	17.8	37.8	17.8	2.0	100.0
Avg. carbon no.	10.3	10.4	9.8	9.6	9.7	10.0
Avg. MW	146	147	137	128	128	139
% Carbon	84.66	84.90	85.71	89.79	91.08	86.21
% Hydrogen	15.34	15.10	14.29	10.21	8.92	13.79

average MS, etc., as well as proprietary process-related parameters, are based on this summary.

Overall Evaluation

Evaluation of the expert program is still in progress, as is its further development. This process essentially involves answering a number of key questions, including: Are identifications correct? How reliable is the program? How does it compare with conventional MS analyses of the total sample? How much time is saved?

Answers to these questions show that the program is indeed reliable and saves a great deal of human expert time.

Overall reliability is monitored by comparing the identification arrived at by the program with those by a human expert on the same sample. These studies show a 95% plus agreement if the number of components is considered, and 98 to 99% plus agreement if the total percent by weight of the identified components is considered (Table 6).

In addition, the expert program detects minor or trace components that are very difficult to be seen by a human expert.

There is also a good agreement with conventional MS and GC analyses, as illustrated by the data reported on Table 7. The GC/MS data given are the sums of the individual components, sorted to correspond with a conventional type analysis obtained on a CEC 103-C mass spectrometer.

Conclusion and Future Plans

The expert program described in this article has met the initial objectives that were outlined at the inception of the effort. It has been shown to be highly reliable in correctly identifying components in complex hydrocarbon streams, and doing so at a fraction of the time that would be required by a human expert. Its quantitation and sorting capabilities are very useful in providing the industrial user with easily accessible and understandable data, that can be immediately used in decision-making affecting processing and end-product use.

The development effort was also a successful example of university-industry cooperation. The work was carried out jointly by Exxon Research and the University of Houston; one of us, Lan Le, then a graduate student at the University of Houston, worked at Exxon Research in Baytown for a considerable time in the role of "knowledge engineer" and programmer.

TABLE 6—*Evaluation of expert program: comparison with human expert.*

Item	Data, Results
Number of peaks	139
Number of components	320 (1 to 4 components per peak)
Human versus "program"	
+ agreed identification	166
+ disagreed on	2
+ additional minor components found only by program	152
Time required by human interpretation, quantitation	20+ h
Time required by program	~1 h or less

TABLE 7—*Expert Program calculations agree well with other analytical methods.*

Component Type	Percent by Weight		
	GC/MS	MS	GC
OVERALL TYPE DISTRIBUTION			
Paraffins	54.27	57.48	
Normal	20.86	...	27.16
ISO	33.41	...	...
1-Ring cycloparaffins	35.08	33.15	...
2-Ring cycloparaffins	7.54	8.91	...
3-Ring cycloparaffins	0.05	...	
Alkylbenzenes	0.15	0.46	
Unidentified	2.91		
N-PARAFFIN DISTRIBUTION			
C_9	0.19	...	0.47
C_{10}	6.83	...	10.63
C_{11}	13.43	...	15.80
C_{12}	0.38	...	0.25
C_{13}	0.01	...	0.01
C_{14}	0.01	...	...
C_{15}	0.01	...	...
	...	...	...
	20.86	...	27.16

Future plans include further refining of the rules, including DENDRAL type structure-spectra correlations; sorting of some of the isomer sets according to known retention time indices; increasing the number of type analyses available to the program; using low electron voltage spectra to facilitate identification and quantitation of aromatic components, etc.

Applicability of the approach to several types of commercial GC/MS/DS systems will also be investigated.

Acknowledgment

The effort discussed in this article would not have been possible without the important contributions and helpful advice from several colleagues. These include: Professor S. Deming, University of Houston; D. L. Larsen, Exxon Research & Engineering Co.; P. K. Bradley, Exxon Chemical Co.; Hazel Hill, Exxon Research & Engineering Co.; Patricia Sharp, Exxon Research & Engineering Co.; and Debra Richards, Exxon Research & Engineering Co.

Ronald D. Grigsby[1]

Processing of High Resolution Mass Spectral Data by Use of Kendrick Masses in a Rectangular Array

REFERENCE: Grigsby, R. D., **"Processing of High Resolution Mass Spectral Data by Use of Kendrick Masses in a Rectangular Array,"** *Novel Techniques in Fossil Fuel Mass Spectrometry, ASTM STP 1019,* T. R. Ashe and K. V. Wood, Eds., American Society for Testing and Materials, Philadelphia, 1989, pp. 172–193.

ABSTRACT: Computer programs using the Kendrick scale for calculating masses are described for rapidly assigning elemental formulas to ions in high resolution mass spectra. Output from the programs, based on nominal mass **Z** series, is displayed in the form of a rectangular array having 14 columns and a variable number of rows, depending on the number of peaks in a spectrum. This layout permits formulas for the major ions to be identified almost as fast as the output is obtained from the printer. The array is especially useful for displaying the spectra of complex petroleum fractions, where molecular ions for one thousand or more components may be detectable. All programs except one are written in FORTRAN 77 and are available from the author.

The utility of the programs is illustrated with output from the mass spectrum of a fraction known to contain one-ring aromatic neutrals separated from the 200 to 425°C distillate of the Cerro Negro (Eastern Venezuela) heavy petroleum. The fraction was found to consist mostly of hydrocarbons with **Z** numbers ranging from −6 through −16, where **Z** is defined by the formula C_nH_{2n+z}. Homologs in the −8 and −10 **Z** series were the most abundant. Only traces of oxygen- and nitrogen-containing compounds were found, but sulfur compounds having formulas $C_nH_{2n+z}S$ with $Z \leq -4$ were found in significant quantities. These types of compounds are in agreement with those anticipated from the HPLC method used in the fractionation of the distillate.

KEY WORDS: fossil fuels, mass spectrometry, petroleum analysis, data processing

High resolution mass spectrometry (HRMS) is one of the most versatile methods available for analyzing petroleum and other sources of hydrocarbon fuels. Because the technique provides elemental compositions and quantitative data on compounds present in complex mixtures, it has considerable value for monitoring refining processes where lower quality feedstocks are being utilized.

Ordinarily, HRMS is not used to analyze crude petroleum, but it is employed instead with separation techniques to analyze fractions produced from the original crude [*1*]. Even when the petroleum is broken down according to compound type, the resultant fractions may still contain well over one thousand detectable components. Owing to this complexity, the identification of specific compounds is usually not attempted; instead, the method is used to classify the compounds according to type, for example, alkylbenzenes.

Data produced by HRMS consist of masses accurate within a few parts per million (typ-

[1] Senior chemist, IIT Research Institute, National Institute for Petroleum and Energy Research, Bartlesville, OK 74005.

ically 10 ppm or 5 mmu at mass 500) and intensities falling within a dynamic range of 10 000 or more. Because elemental compositions of the molecular ions are of greatest interest, the intensities of fragment ions are usually suppressed by using low-energy electron ionization to generate the ions.

When the spectrum of a single petroleum fraction (for example, the aromatic neutrals) contains over one thousand molecular ions, the task of assigning elemental compositions to these ions becomes a formidable one that may consume so much time as to render the cost of the analysis prohibitive. Any program used to process data from the spectrum of a mixture containing a multitude of components is inefficient in terms of time and cost if it produces a list of formulas for each ion that must be examined manually to select the best choice. This selection process may take several days or sometimes more than a week, depending on the complexity of the spectrum and the amount of information that must be obtained from it. Clearly, any means by which the effort can be reduced will be beneficial to the laboratory providing the analysis. Even when efficient computer programs are available to generate elemental compositions and theoretical masses for comparison with experimental masses, one or more criteria besides the limit on experimental error oftentimes must be used to select a particular formula from those produced by the program. If the data are to be used in a quantitative analysis of the fraction, the intensities of the ions—for which formulas have been assigned—may have to be entered into another program (perhaps manually) that calculates the quantitative results. Thus, the slow step in processing high resolution mass spectral data for a complex mixture is the assignment of the most probable formula to each significant ion, and any method that can reduce the time required for the assignments will result in a corresponding reduction in the cost of the analysis.

All manufacturers of high resolution mass spectrometers provide programs for assigning elemental formulas to ions in the spectra. These programs may or may not be adequate for processing the data, depending on the particular application. To reduce the time required to process the data from petroleum fractions, a novel method for formula assignment has been investigated in our laboratory. This effort has led to the development of computer programs that can partially assign formulas to the ions and produce output that is easily edited to complete the assignments. These programs are described along with output resulting from their execution on a typical data set obtained from the mass-spectral analysis of a petroleum fraction.

The programs have been developed around the concept of a repeating unit that exists in compounds that can be classified into homologous series. For hydrocarbons and other compounds occurring in petroleum, this unit is CH_2 or methylene. Over two decades ago, Kendrick showed that the interpretation of high resolution mass spectra of hydrocarbon-derived mixtures could be simplified if the masses of the ions were based on CH_2 instead of ^{12}C [*2*]. Ions are classified into homologous series whose members have masses differing by exactly 14 amu, that is, each member has exactly the same mass defect. Recently, the Kendrick mass scale has been generalized to include CF_2, CCl_2, and CBr_2, as additional repeating units, and programs for producing elemental formulas based on these and the CH_2 unit are described in the literature [*3,4*].

Experimental

The distillate of the Cerro Negro crude has been fractionated into acids, bases, and neutrals with subfractionation of each according to aromatic ring number. The subfraction considered here consists predominantly of one-ring aromatics separated from the neutrals fraction [*1*].

A Kratos MS-50 high resolution mass spectrometer (Kratos Analytical Instruments,

Manchester, UK) was used to record the spectrum of the fraction at a resolving power of 40 000 (10% valley definition). The sample was introduced through an all-glass expansion volume inlet (R. J. Brunfeldt Co., Bartlesville, OK) at 300°C, and ions were generated by electron ionization at a nominal energy of 10 eV.

Discussion

The difference between Kendrick mass **(KM)** and ordinary ^{12}C mass is seen from a comparison of masses defined by the two scales. In Table 1, elemental compositions and masses are listed for several C_nH_{2n} combinations. With the mass of carbon defined exactly as 12 amu, the mass defects of the members (defined as the difference between the mass of a species and its mass number [*3*]) increase with increasing elemental complexity. This causes the integer mass to be greater than the mass number **(MN),** that is, the sum of the numbers of protons and neutrons, when the mass defect exceeds 1.0000. Because the mass defect is not constant over a homologous series and because the integer mass is not always the same as the mass number, unnecessary complications arise in the interpretation of the spectrum. These can be avoided by replacing the ^{12}C mass scale with the Kendrick scale. Thus, the integer mass becomes identical to the mass number for all members of the C_nH_{2n} series, that is, the mass defect is zero over the series. For other series, the mass defects are not zero although they are constant. The integer masses are not generally equal to the mass numbers even though the latter always can be obtained by subtracting the Kendrick mass defects from the theoretical or calculated Kendrick masses.

The Kendrick mass scale is readily incorporated into computer programs for processing mass spectral data with output in the form of a rectangular array. This means of displaying a mass spectrum is consistent with the notion of nominal mass **Z** series **(NMZ),** a concept that has been used for many years by researchers in the petroleum industry [*3,5–8*]. The array has 14 columns and a variable number of rows, depending on the length of the spectrum, and adjacent elements in each column differ by 14 amu, corresponding to the mass of CH_2. For high-resolution mass spectra of homologs containing C, H, and a heteroatom or heteroatomic group, Y, the array classifies ions into specific **Z** series **(SPZ),** where **Z** is defined by the formula $C_nH_{2n+z}Y$.

TABLE 1—*Elemental compositions and masses for C_nH_{2n} homologs with mass defined by ^{12}C and Kendrick scales.*

Elemental Composition	^{12}C Mass	Kendrick Mass
CH_2	14.0157	14.0000
C_2H_4	28.0313	28.0000
C_3H_6	42.0470	42.0000
...	...	...
...	...	...
...	...	...
$C_{40}H_{80}$	560.6260	560.0000
...	...	...
...	...	...
...	...	...
$C_{100}H_{200}$	1401.5650	1400.0000
...	...	...
...	...	...
...	...	...

Table 2 presents the unedited partial hydrocarbon **(CH)** output for the spectrum of one-ring aromatic neutrals separated from the 200 to 425°C distillate of the Cerro Negro heavy petroleum. Two arrays are shown: in the first (Table 2*a*), $-11 \leq$ **SPZ** $\leq +2$ and in the second (Table 2*b*) $-25 \leq$ **SPZ** ≤ -12. If a peak is present at a given mass number, its intensity and parts-per-million error (ppm) are placed just below the mass number in the array, and the intensities of all peaks in the spectrum, including those having no assignments, are normalized to make their sum equal 100 000. The major compound types are found in the -8 and -10 **SPZ** series (first array), as seen by the sums of the intensities given at the ends of the corresponding columns. If alkane molecular ions were present, they would fall into the column for which **SPZ** $= +2$. Naphthalene molecular ions, on the other hand, are classified under the column heading **SPZ** $= -12$ in the second array. All unassigned peaks are classified by **NMZ** series at the end of the output (Table 2*c*). This arrangement is convenient for editing both parts of the output, as explained later.

Major fragment ions fall into the column just to the left of the one containing the molecular ions; for example, **SPZ** $= +1$ for alkane fragments and -7 for fragments from alkylbenzenes. The output could be simplified by omitting fragment-ion intensities; however, they are useful for identifying compounds and are required in making isotope corrections, if these are necessary.

Because each element of an array column differs exactly by 14 amu from the adjacent elements, the Kendrick mass is automatically specified for any ion having a given mass number once the mass of any other ion in the series is identified. Furthermore, the mass of any ion of a given mass number in any row of the array is specified because masses corresponding to adjacent row elements differ exactly by the Kendrick mass of the hydrogen atom, 1.006700 amu. The Kendrick mass factor **(KMF)** is 14 divided by the mass of CH_2 as defined by the ^{12}C scale. Thus, any Kendrick mass can be calculated from the corresponding ^{12}C mass simply by multiplying the latter by **KMF** $= 0.9988834$.

The Kendrick mass corresponding to any mass number in an array is obtained by adding the Kendrick mass defect **(KMD)** to the mass number. This follows from the definition of **KMD** as the difference between the Kendrick mass and the mass number [*3*]. Thus, **KMD** < 0 when **KM** $<$ **MN** and **KMD** ≥ 0 when **KM** $\geq$ **MN.** For example, benzene has a **KMD** of $77.9598 - 78 = -0.040200$ amu, or conversely, the Kendrick mass of benzene is $78 - 0.040200 = 77.9598$ amu. Between columns, the difference in **KMDs** is constant and given by the **KMD** of the hydrogen atom, that is, 0.006700 amu. With these thoughts in mind, it becomes a simple matter to calculate the Kendrick mass of any ion whose mass number is given in the array by adding **KMD** + 1 to one less than the mass number for the ion. The Kendrick mass of the decane molecular ion, for example, is $142 - 1 + 1.0134 = 142.0134$ amu. Values of **KMD** + 1 are given in a row at the top of the array. For arrays other than those for hydrocarbons, the Kendrick masses are calculated in the same manner, except that **KMD** will not be zero when **SPZ** = 0.

To facilitate identification of the elemental composition corresponding to any mass number in the array, the carbon number for each row is given on the right-hand side under the column heading "N." The number of hydrogens for any mass number in the array can be obtained by starting with the maximum number on the right ("2N+2" for the first hydrocarbon array, Table 2*a*) and subtracting one for each member of the row when counting from right to left.

Nominal mass **Z** series numbers are identical to the hydrocarbon **SPZ** numbers -11 through $+2$. However, **NMZ** series are also useful for describing ions in the spectra of heteroatomic compounds even though the **NMZ** and **SPZ** numbers do not necessarily coincide. The **NMZ** number depends only on the mass number and not on elemental composition. This leads to the definition of the nominal mass **Z** series number as the remainder

TABLE 2a—*Partial mass spectrum of one-ring aromatic neutrals separated from the 200 to 425°C distillate of the Cerro Negro (Eastern Venezuela) heavy petroleum. Specific Z numbers from −11 through +2.*

NAME: RSPCNM. 8/12/86. CN 200-425 DEG C 1-RING COMPDS. RUN 1. SCAN NO.: 1
MAXIMUM PPM ERROR = 6.0 TOTAL ION CURRENT = 0.3152E+07 ELEMENTAL COMPOSITION: C(N)H(2N+Z)

NOMINAL MASS Z NUMBERS

−11	−10	−9	−8	−7	−6	−5	−4	−3	−2	−1	0	1	2		
KENDRICK MASS DEFECT + 1.0000															
0.9263	0.9330	0.9397	0.9464	0.9531	0.9598	0.9665	0.9732	0.9799	0.9866	0.9933	1.0000	1.0067	1.0134		
SPECIFIC Z NUMBERS															
−11	−10	−9	−8	−7	−6	−5	−4	−3	−2	−1	0	1	2	N	2N+2
MASS/INTENSITY/PPM ERROR															
									12.0	13.0	14.0	15.0	16.0	1	4
	...	...	...	...	...	...	...	...	...	...	...	...	...	...	...
	...	...	...	...	...	...	...	...	...	...	...	...	...		
	...	...	...	...	...	...	...	...	...	...	...	...	...		
73.0	74.0	75.0	76.0	77.0	78.0	79.0	80.0	81.0	82.0	83.0	84.0	85.0	86.0	6	14
								81.2		119.0					
								2.4		1.7					
87.0	88.0	89.0	90.0	91.0	92.0	93.0	94.0	95.0	96.0	97.0	98.0	99.0	100.0	7	16
					111.0				23.6			24.0			
					1.1				−0.7			4.4			
101.0	102.0	103.0	104.0	105.0	106.0	107.0	108.0	109.0	110.0	111.0	112.0	113.0	114.0	8	18
			63.9	206.1			46.9	182.5		211.9		29.2			
			−1.0	2.1			1.0	−0.2		5.8		−4.7			
115.0	116.0	117.0	118.0	119.0	120.0	121.0	122.0	123.0	124.0	125.0	126.0	127.0	128.0	9	20
	62.0	96.5	273.3									29.0			
	4.5	−0.1	−5.1									0.1			
129.0	130.0	131.0	132.0	133.0	134.0	135.0	136.0	137.0	138.0	139.0	140.0	141.0	142.0	10	22
41.7	271.1				1038.2		45.2	115.0		54.1					
2.1	−3.5				−5.2		−1.5	−3.2		−3.4					
143.0	144.0	145.0	146.0	147.0	148.0	149.0	150.0	151.0	152.0	153.0	154.0	155.0	156.0	11	24

149.8 −1.2	353.3 0.0	1221.7 1.2	450.2 0.8	623.7 3.5	482.2 1.5	64.7 0.3	39.1 1.3	56.0 1.9		33.9 −3.8					
157.0 649.6 −4.5	158.0 497.5 0.9	159.0 1378.6 0.6	160.0 455.6 1.5	161.0 412.7 2.2	162.0 347.7 1.2	163.0 70.4 −3.0	164.0	165.0 62.7 1.1	166.0	167.0	168.0	169.0	170.0	12	26
171.0	172.0 355.4 −5.6	173.0 1260.9 −1.3	174.0 783.6 −1.0	175.0 301.3 −2.3	176.0 304.8 −1.6	177.0 76.7 1.2	178.0	179.0 47.1 2.9	180.0	181.0	182.0	183.0	184.0	13	28
185.0 804.5 −3.8	186.0 643.6 −1.8	187.0 1014.9 1.9	188.0 1093.1 2.8	189.0 232.4 0.6	190.0 379.8 −0.6	191.0 35.8 2.2	192.0	193.0 45.4 −2.8	194.0	195.0	196.0	197.0	198.0	14	30
199.0 836.2 −2.4	200.0 790.4 0.8	201.0 646.8 0.2	202.0 1403.2 2.4	203.0 186.3 0.5	204.0 471.2 2.0	205.0 19.9 6.0	206.0	207.0 18.3 4.2	208.0	209.0	210.0	211.0	212.0	15	32
213.0 793.3 0.2	214.0 1032.6 3.8	215.0	216.0 946.7 5.2	217.0	218.0 462.2 5.0	219.0	220.0	221.0	222.0	223.0	224.0	225.0	226.0	16	34
227.0	228.0	229.0 503.8 −5.0	230.0 994.6 −1.6	231.0	232.0 430.3 −0.5	233.0	234.0	235.0	236.0	237.0	238.0	239.0	240.0	17	36
241.0	242.0 1070.5 −4.5	243.0 315.3 −3.5	244.0	245.0 179.2 2.1	246.0 430.2 −5.6	247.0	248.0	249.0	250.0	251.0	252.0	253.0	254.0	18	38
255.0	256.0	257.0	258.0	259.0	260.0	261.0 55.8 3.3	262.0	263.0	264.0	265.0	266.0	267.0	268.0	19	40
269.0 451.6 0.1	270.0 799.7 0.9	271.0	272.0 678.9 1.2	273.0 28.6 −0.7	274.0 291.7 1.9	275.0	276.0	277.0	278.0	279.0	280.0	281.0	282.0	20	42
... ...	... ...	... ...	... ...	... ...	... ...	... ...	... ...	... ...	... ...	... ...	... ...	... ...	... ...	...	...
							COLUMN SUMS:								
5218.4	10142.2	7046.8	10578.0	2316.8	5712.9	334.8	131.1	608.2	23.6	418.9	0.0	82.2	0.0		

TABLE 2*b*—*Specific Z numbers from −25 through −12.*

NOMINAL MASS Z NUMBERS															
−11	−10	−9	−8	−7	−6	−5	−4	−3	−2	−1	0	1	2		
KENDRICK MASS DEFECT + 1.0000															
0.8325	0.8392	0.8459	0.8526	0.8593	0.8660	0.8727	0.8794	0.8861	0.8928	0.8995	0.9062	0.9129	0.9196		
SPECIFIC Z NUMBERS															
−25	−24	−23	−22	−21	−20	−19	−18	−17	−16	−15	−14	−13	−12	N	2N−12
MASS/INTENSITY/PPM ERROR															
													72.0	6	0
						...	...	...	...	...	...	...	...	...	...
						...	...	...	...	...	...	...	...		
						...	...	...	...	...	...	...	...		
					120.0	121.0	122.0	123.0	124.0	125.0	126.0	127.0	128.0	10	8
												27.3	8.6		
												1.6	−2.0		
...	...	...	...	...	...	...	...	...	...	...	...	...	...	...	...
...	...	...	...	...	...	...	...	...	...	...	...	...	...		
...	...	...	...	...	...	...	...	...	...	...	...	...	...		
185.0	186.0	187.0	188.0	189.0	190.0	191.0	192.0	193.0	194.0	195.0	196.0	197.0	198.0	15	18
											39.6	103.7	210.2		
											1.0	2.1	−0.8		

199.0	200.0	201.0	202.0	203.0	204.0	205.0	206.0	207.0	208.0 10.5 2.8	209.0	210.0 118.8 −0.9	211.0	212.0 265.1 −0.1	16	20
213.0	214.0	215.0	216.0	217.0	218.0	219.0	220.0	221.0	222.0 7.2 4.6	223.0 55.0 0.8	224.0	225.0	226.0	17	22
227.0	228.0	229.0	230.0	231.0	232.0	233.0	234.0	235.0	236.0	237.0	238.0 202.7 −1.3	239.0 389.4 −5.9	240.0	18	24
241.0	242.0	243.0	244.0	245.0	246.0	247.0	248.0	249.0	250.0	251.0	252.0 219.9 −1.8	253.0 761.8 −1.6	254.0 529.0 −0.9	19	26
255.0	256.0	257.0	258.0	259.0	260.0	261.0	262.0	263.0	264.0	265.0	266.0	267.0	268.0 697.9 −2.8	20	28
269.0	270.0	271.0	272.0	273.0	274.0 24.9 4.7	275.0 21.1 3.9	276.0	277.0 23.3 1.0	278.0 312.3 −2.1	279.0 118.0 −0.9	280.0 206.1 −2.0	281.0 303.8 3.5	282.0 627.9 1.8	21	30
...	...	...	...	...	...	...	...	...	...	...	...	...	...	...	...
...	...	...	...	...	...	...	...	...	...	...	...	...	...		
...	...	...	...	...	...	...	...	...	...	...	...	...	...		
COLUMN SUMS															
0.0	64.5	25.2	193.4	129.4	314.8	331.3	523.0	605.1	2115.3	996.5	3205.6	2535.0	5663.6		

TABLE 2*c*—*Unassigned peaks classified by nominal mass Z series.*

NAME: RSPCNM. 8/12/86. CN 200-425 DEG C 1-RING COMPDS. RUN 1. SCAN NO.: 1
MAXIMUM PPM ERROR = 6.0 TOTAL ION CURRENT = 0.3152E+07 ELEMENTAL COMPOSITION: C(N)H(2N+Z)
ORIGINAL NO. OF MASSES = 671 NO. OF MASSES ASSIGNED = 256 PERCENT OF TOTAL ION CURRENT ASSIGNED = 59.3
UNASSIGNED MASSES AND INTENSITIES:

NOMINAL MASS Z NUMBERS

−11	−10	−9	−8	−7	−6	−5	−4	−3	−2	−1	0	1	2
100.929 86.1	115.936 14.0	117.015 20.2	131.942 36.4	118.950 17.4	119.924 19.2	120.963 117.3	93.972 11.8	94.979 124.5	109.986 31.4	96.933 11.8	97.989 34.4	85.006 78.6	113.923 26.8
114.930 178.2	143.928 19.0	130.937 10.8	131.948 467.4	118.954 590.2	119.956 20.0	120.967 34.7	121.974 43.9	108.924 241.3	123.988 36.0	96.934 12.9	97.995 12.1	126.894 7.7	127.769 12.8
128.930 270.6	157.928 38.7	130.941 587.6	145.942 115.3	132.949 37.1	119.958 13.0	134.956 213.4	135.962 1384.2	122.981 226.1	137.897 24.2	96.994 255.6	111.910 58.4	126.916 37.4	127.923 37.1
142.930 226.0	171.928 58.5	144.935 18.0	159.941 187.2	132.954 988.4	119.961 1174.5	134.962 62.4	135.975 8.5	150.889 11.6	151.896 17.1	124.904 103.0	125.910 79.8	140.916 39.8	141.924 21.1
... ...	... ...	... ...	... ...	... ...	... ...	... ...	... ...	... ...	... ...	... ...	... ...	... ...	... ...
268.910 10.1	367.842 30.2	256.942 207.0	327.857 98.4	216.947 136.0	259.874 23.6	218.961 78.2	289.882 113.8	262.890 148.7	291.890 47.1	264.903 239.6	265.903 7.1	238.908 19.3	309.915 27.4

282.835	381.844	270.849	327.910	216.952	259.962	232.962	303.883	276.889	291.895	278.894	265.910	252.670	351.911
14.6	22.6	7.3	20.2	153.6	391.3	60.2	99.1	41.1	230.8	25.6	295.3	38.7	33.4
...	...	...	...	...	...	...	...	...	...	...	...	...	...
...	...	...	...	...	...	...	...	...	...	...	...	...	...
549.024		396.936	425.858	286.949	343.870	302.964	500.072	626.842		545.017	545.923	406.908	
8.4		40.6	55.2	121.4	177.1	17.8	17.8	12.0		15.5	19.0	38.1	
		410.937	439.182	300.949	357.958	316.963						406.921	
		90.9	18.7	90.5	53.0	14.7						13.1	
		424.935		314.863	371.869	330.961						420.910	
		22.3		29.6	152.9	28.5						22.5	
		438.936		314.950	385.870	344.962						420.916	
		5.4		116.3	125.8	43.5						39.6	
		495.073		328.857	399.869	386.601						448.917	
		5.2		13.7	146.0	13.1						22.2	
				328.948	413.871	386.876						533.016	
				61.5	66.6	25.5						40.0	
				342.949	427.873	414.876							
				80.5	61.3	19.1							
...	...	...	...	...	...	...	...	...	...	...	...	...	...
...	...	...	...	...	...	...	...	...	...	...	...	...	...

obtained when the mass number is divided by 14. According to conventional terminology used in petroleum mass spectrometry, the remainder is forced to fall between -11 and $+2$ [*3,6*]. This is equivalent to the statement that 14 is to be subtracted from the mass number n times such that the remainder falls between -11 and $+2$, that is, $-11 \leq$ **NMZ** $\leq +2$. Classification of a number of compound types by **NMZ** series is shown in Table 3. A much more extensive discussion is given in Ref *5*.

As a further illustration of the difference between **NMZ** and **SPZ** series, the **CHO_2** array presented in Table 4 shows that molecular ions for aliphatic carboxylic acids fall into the column with **NMZ** = -10. On the other hand, **SPZ** = 0, because it is defined by the formula $C_nH_{2n+z}O_2$. No **CHO_2** peaks appear in the output because none were detected in the spectrum of the one-ring aromatic neutrals.

Six programs, run sequentially, are required to process raw mass/intensity data into the array format. These programs are named **SPECTR, XFER, READTAP, PROB, CHY,** and **CH13CY.** All programs except **XFER** (Data General Software) are written in FORTRAN 77 and are available from the author. All programs except **SPECTR** and **XFER** run on a Digital Equipment Corporation MicroVAX-II computer. The other two execute on a Data General NOVA/4, which is a component part of the Kratos data acquisition system.

The relationship between programs and files is shown in Fig. 1. Working with a file containing high resolution mass/intensity data acquired by the data-acquisition system, programs **SPECTR** and **XFER** restructure the data into a tape format compatible with the input requirements of the MicroVAX-II. Transfer of data on tape to this computer is handled by program **READTAP,** which places the data in file **READTAP.OUT.** From this file, program **PROB** accepts the data, consisting of ^{12}C masses and intensities in order of decreasing mass and converts the masses to the Kendrick scale. The data are reordered by increasing mass and processed to separate the reference peaks (originating from a calibration mixture run simultaneously with the sample) from the sample peaks. Each mass is matched against a set of reference masses contained in file **REF.DAT,** and a peak is identified as a reference if its mass is within a specified parts-per-million range of the mass of the reference ion. The maximum allowed error is usually set at 10 ppm or less, depending on the resolving power at which the spectra were acquired and on other factors.

Two output files are generated by program **PROB:** one, **PROB1.OUT** contains the identified reference peaks; and the other, **PROB.OUT** consists of all remaining peaks, which are assumed to arise from the sample. The latter file is renamed and used as input to program **CHY** that identifies ions belonging to specified homologous series, for example, C_nH_{2n+z} or $C_nH_{2n+z}O$. Separate output files, **CHY.OUT,** are generated for each range of **Z** numbers, for example, $+2$ through -53, corresponding to the specified elemental combination.

When program **CHY** is run, the proper array is automatically set up from information requested at the terminal (including the most positive and most negative **SPZ** numbers). The ^{12}C mass of the heteroatom or heteroatomic group is entered, for example, 15.994915 or 31.989830 for O or O_2, respectively. For hydrocarbons, the requests are ignored because there is no heteroatomic group. The program then calculates **KMD** + 1 for the group and assigns this value to the array column for which **SPZ** = 0. The **KMD** + 1 values are calculated for all other columns, and the mass numbers are placed at the proper locations in the array. Intensities and ppm errors for the experimental masses are placed below the corresponding mass numbers according to an algorithm to be described. The array is not limited to any particular isotope of a heteroatom. As an illustration, the first array for **$CH^{34}S$** combinations is given in Table 5. No peaks appear in the partial array shown because all ion intensities corresponding to the mass numbers in Table 5 were below threshold.

TABLE 3—*Classification of compound types by nominal mass Z series.*

NOMINAL MASS Z NUMBERS													
−11	−10	−9	−8	−7	−6	−5	−4	−3	−2	−1	0	1	2
MASS													
3	4	5	6	7	8	9	10	11	12	13	14	15	16
17	18	19	20	21	22	23	24	25	26	27	28	29	30
31	32	33	34	35	36	37	38	39	40	41	42	43	44
45	46	47	48	49	50	51	52	53	54	55	56	57	58
59	60	61	62	63	64	65	66	67	68	69	70	71	72
73	74	75	76	77	78	79	80	81	82	83	84	85	86
87	88	89	90	91	92	93	94	95	96	97	98	99	100
101	102	103	104	105	106	107	108	109	110	111	112	113	114
...	...	...	...	...	...	...	...	...	...	...	...	...	...
...	...	...	...	...	...	...	...	...	...	...	...	...	...
...	...	...	...	...	...	...	...	...	...	...	...	...	...
POSSIBLE COMPOUND TYPES													
	C_nH_{2n-10}		C_nH_{2n-8}		C_nH_{2n-6}		C_nH_{2n-4}		C_nH_{2n-2}		C_nH_{2n}		C_nH_{2n+2}
	C_nH_{2n-24}		C_nH_{2n-22}		C_nH_{2n-20}		C_nH_{2n-18}		C_nH_{2n-16}		C_nH_{2n-14}		C_nH_{2n-12}
	$C_nH_{2n+2}O$		$C_nH_{2n-10}O$		$C_nH_{2n-8}O$		$C_nH_{2n-6}O$		$C_nH_{2n-4}O$		$C_nH_{2n-2}O$		$C_nH_{2n}O$
	$C_nH_{2n}O_2$		$C_nH_{2n+2}O_2$		$C_nH_{2n-10}O_2$		$C_nH_{2n-8}O_2$		$C_nH_{2n-6}O_2$		$C_nH_{2n-4}O_2$		$C_nH_{2n-2}O_2$
$C_nH_{2n+3}N$		$C_nH_{2n-9}N$		$C_nH_{2n-7}N$		$C_nH_{2n-5}N$		$C_nH_{2n-3}N$		$C_nH_{2n-1}N$		$C_nH_{2n+1}N$	
	$C_nH_{2n}S$		$C_nH_{2n+2}S$		$C_nH_{2n-10}S$		$C_nH_{2n-8}S$		$C_nH_{2n-6}S$		$C_nH_{2n-4}S$		$C_nH_{2n-2}S$
...	...	...	...	...	...	...	...	...	...	...	...	...	...
...	...	...	...	...	...	...	...	...	...	...	...	...	...
...	...	...	...	...	...	...	...	...	...	...	...	...	...

TABLE 4—*First array for CHO_2 combinations. Molecular ions for aliphatic carboxylic acids are found in the −10 nominal mass Z series (0 specific Z series) column.*

NAME: RSPCNM. 8/12/86. CN 200-425 DEG C 1-RING COMPDS. RUN 1. SCAN NO.: 1
MAXIMUM PPM ERROR = 6.0 TOTAL ION CURRENT = 0.3152E+07 ELEMENTAL COMPOSITION: C(N)H(2N+Z)O(2)

NOMINAL MASS Z NUMBERS															
−11	−10	−9	−8	−7	−6	−5	−4	−3	−2	−1	0	1	2		
KENDRICK MASS DEFECT + 1.0000															
0.9474	0.9541	0.9608	0.9675												
SPECIFIC Z NUMBERS															
−1	0	1	2											N	2N+2
MASS/INTENSITY/PPM ERROR															
45.0	46.0	47.0	48.0											1	4
59.0	60.0	61.0	62.0											2	6
73.0	74.0	75.0	76.0											3	8
87.0	88.0	89.0	90.0											4	10
101.0	102.0	103.0	104.0											5	12
115.0	116.0	117.0	118.0											6	14
129.0	130.0	131.0	132.0											7	16
143.0	144.0	145.0	146.0											8	18
157.0	158.0	159.0	160.0											9	20
...	...	...	...	...	...	...	...	...	...	...	...	...	...	...	...
...	...	...	...	...	...	...	...	...	...	...	...	...	...		
...	...	...	...	...	...	...	...	...	...	...	...	...	...		

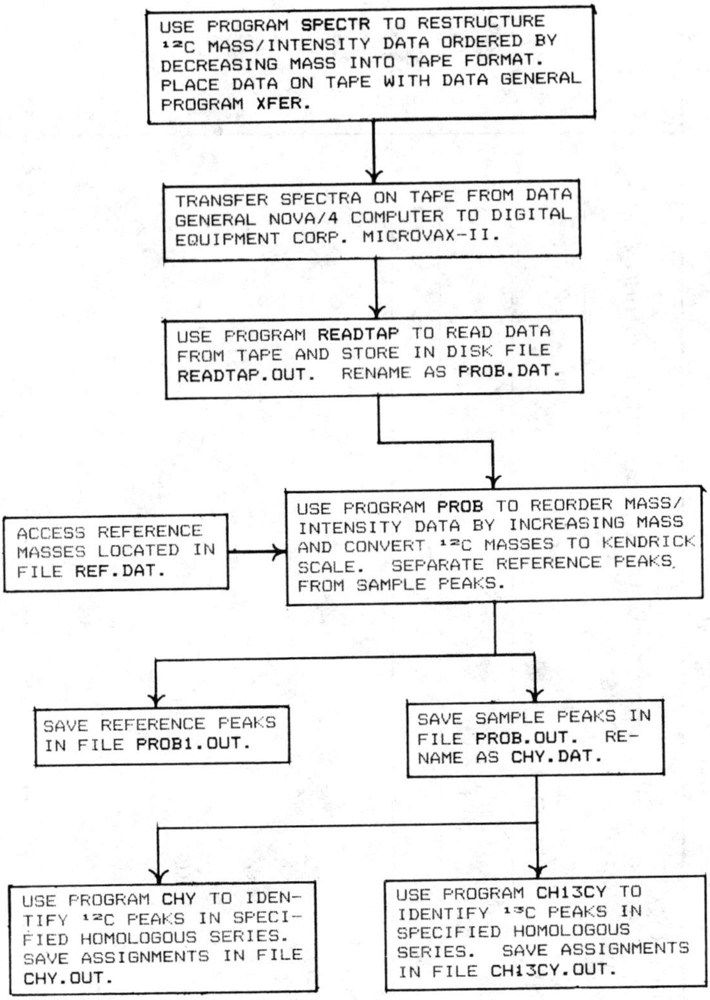

FIG. 1—*Flow chart showing the relationship between programs and files.*

For combinations containing ^{13}C, a separate program, **CH13CY,** is used (Fig. 1). When the ^{12}C mass of the heteroatomic group is entered, the mass of $^{13}CH_2$ (15.019005 on the ^{12}C scale) is added to it, that is, $^{13}CH_2$ is considered to be part of the group. Thus, for **CH^{13}CNO** combinations, **KMD** + 1 = **KMF**(15.019005 + 29.997989) − 45 + 1 = 0.9667 when **SPZ** = 0. Also, when the total number of carbon atoms is calculated for a row, an additional carbon must be added because $^{13}CH_2$ has been removed from the **CH** part and included with the heteroatomic group. Combinations containing more than one ^{13}C would be handled in a similar manner. Table 6 shows the first array set up for **CH^{13}CNO** combinations. No peaks having these combinations were detected in the spectrum of the Cerro Negro fraction.

Once the array is laid out for a particular elemental combination, it is a relatively simple matter to match experimental and theoretical masses and thereby specify the most probable formula for an ion. Although this task can be managed by various algorithms, the one

TABLE 5—*First array for $CH^{34}S$ combinations.*

NAME: RSPCNM. 8/12/86. CN 200-425 DEG C 1-RING COMPDS. RUN 1. SCAN NO.: 1
MAXIMUM PPM ERROR = 6.0 TOTAL ION CURRENT = 0.3152E+07 ELEMENTAL COMPOSITION: C(N)H(2N+Z)34S

NOMINAL MASS Z NUMBERS															
−11	−10	−9	−8	−7	−6	−5	−4	−3	−2	−1	0	1	2		
KENDRICK MASS DEFECT + 1.0000															
0.9098	0.9165	0.9232	0.9299	0.9366	0.9433										
SPECIFIC Z NUMBERS															
−3	−2	−1	0	1	2									N	2N+2
MASS/INTENSITY/PPM ERROR															
	46.0	47.0	48.0	49.0	50.0									1	4
59.0	60.0	61.0	62.0	63.0	64.0									2	6
73.0	74.0	75.0	76.0	77.0	78.0									3	8
87.0	88.0	89.0	90.0	91.0	92.0									4	10
101.0	102.0	103.0	104.0	105.0	106.0									5	12
115.0	116.0	117.0	118.0	119.0	120.0									6	14
129.0	130.0	131.0	132.0	133.0	134.0									7	16
143.0	144.0	145.0	146.0	147.0	148.0									8	18
157.0	158.0	159.0	160.0	161.0	162.0									9	20
...	...	...	...	...	...	...	...	...	...	...	...	...	...	...	...
...	...	...	...	...	...	...	...	...	...	...	...	...	...		
...	...	...	...	...	...	...	...	...	...	...	...	...	...		

TABLE 6—*First array (partial) for $CH^{13}CNO$ combinations.*

NAME: RSPCNM. 8/12/86. CN 200-425 DEG C 1-RING COMPDS. RUN 1. SCAN NO.: 1
MAXIMUM PPM ERROR = 6.0 TOTAL ION CURRENT = 0.3152E+07 ELEMENTAL COMPOSITION: C(N−1)H(2N+Z)13CNO

NOMINAL MASS Z NUMBERS															
−11	−10	−9	−8	−7	−6	−5	−4	−3	−2	−1	0	1	2		
KENDRICK MASS DEFECT + 1.0000															
0.9667	0.9734	0.9801	0.9868												
SPECIFIC Z NUMBERS															
0	1	2	3											N	2N+3
MASS/INTENSITY/PPM ERROR															
59.0	60.0	61.0	62.0											2	7
73.0	74.0	75.0	76.0											3	9
87.0	88.0	89.0	90.0											4	11
101.0	102.0	103.0	104.0											5	13
115.0	116.0	117.0	118.0											6	15
129.0	130.0	131.0	132.0											7	17
143.0	144.0	145.0	146.0											8	19
157.0	158.0	159.0	160.0											9	21
171.0	172.0	173.0	174.0											10	23
...	...	...	...	...	...	...	...	...	...	...	...	...	...	...	...
...	...	...	...	...	...	...	...	...	...	...	...	...	...		
...	...	...	...	...	...	...	...	...	...	...	...	...	...		

used by programs **CHY** and **CH13CY** selects the best fit within a specified parts-per-million error, E_{ppm}, entered from the terminal when the program is run. In this way, the experimental mass giving the smallest absolute ppm error within the specified range is selected. The associated intensity and ppm error are then placed in the array just below the corresponding mass number. All rejected mass/intensity values are saved for possible assignment in arrays having more negative **SPZ** numbers, as determined from the most negative **SPZ** number entered from the terminal. Any unassigned masses are printed along with their intensities in nominal mass **Z** series at the end of the output, as seen in Table 2*c*.

Further use of the array programs to process high resolution mass spectra is illustrated with a more detailed discussion of the data recorded for the fraction separated from the 200 to 425°C distillate of the Cerro Negro heavy petroleum.

Program **CHY** is run for hydrocarbon homologs first, and, after examination of the output, rerun for heteroatomic combinations. For petroleum fractions, **CHO, CHN,** and **CHS** combinations are always included. Then, additional combinations may be added, as suggested by the output. For example, strong **CHO** peaks may suggest the presence of $\mathbf{CHO_2}$ and $\mathbf{CHO_3}$ peaks, as well. In this way, output for selected elemental compositions is produced.

A principal advantage of displaying a spectrum in the array format is that the major compound types can be immediately identified. For example, by looking at the **CH** output from the Cerro Negro fraction (Table 2), it is evident from the assignments in the columns and from the intensity sums at the ends of the columns that the major **CH** compound types range from C_nH_{2n-6} through C_nH_{2n-16}. The most abundant types are C_nH_{2n-8} (indans and tetralins) and C_nH_{2n-10} (dinaphthenobenzenes) as shown by the -8 and -10 **SPZ** column sums of 10578 and 10142 divisions per 100 000 divisions, respectively. These major compound types are in agreement with the one-ring aromatics predicted by the separation scheme [*1*].

The capability of the programs for making proper assignments and the procedure for editing the output to complete the assignments are illustrated by examining the **CH** output in more detail. Whether an assignment is made depends on the specified maximum ppm error, which was set at 6.0 ppm when the program was run. Excluding the reference peaks, which appear in the **PROB1.OUT** file from program **PROB** (not shown), all peaks not assigned will be found classified by nominal mass **Z** series at the end of the array output (Table 2*c*). Therefore, when a peak is missing from a series, the unassigned list or the list of reference masses is searched for the peak, and, if found, it is included with the series. This editing process consists of first writing the intensity of the peak in by hand under the proper mass number and then modifying the **CHY** (or **CH13CY**) input file to force the assignment, that is, by changing the experimental mass to equal the theoretical mass. Methods in artificial intelligence are being explored to speed up editing, but their development is not far enough advanced to include with the present discussion.

To illustrate how the assignments for a series are edited, the alkylbenzene series is considered in detail. As seen in the first array of Table 2, C_nH_{2n-6} peaks appear from Mass 92 through 274 with the exception of "holes" at Mass 106, 120, and 260. The first member of the alkylbenzene series is benzene, itself, having a Kendrick m/z value of 77.9598 (78 − 1 + 0.9598). This peak is missing because it was used as a reference. It is found in the **PROB1.OUT** file from program **PROB** (not shown) with a raw intensity of 16067 div. The normalized intensity of the peak can be calculated by dividing the total ion current **(TIC)** for the sample peaks into 16067 and multiplying by 1.0×10^5. The **TIC,** 0.3152×10^7, is found at the top of the first page of the output and again at the top of the unassigned list. Thus, $I_{78} = 509.7$. Similarly, m/z 105.9598, C_8H_{10}, is missing because it too was used as a reference mass. Its raw intensity is 61630 div., from which a normalized intensity of 1955.3 is calculated. The missing peak at m/z 119.9598 is not part of the reference file, but is found instead in the list of unassigned peaks in the −6 **NMZ** column at *m/z* 119.961 (Table 2*c*). Its normalized intensity is 1174.5. This peak was omitted from the series because the experimental error of −10.0 ppm is greater than the 6.0 ppm limit set when the program was run. (Note that a negative ppm error means the experimental mass is greater than the theoretical mass.) Although not shown in Table 2*a,* only one other peak is missing from the series up to *m/z* 399.9598. Because peaks above and below Mass 260 in the series have significant intensities (430.2 and 291.7 for *m/z* 245.9598 and 273.9598, respectively), an ion at *m/z* 259.9598 is expected. Examination of the unassigned list (Table 2*c*) shows that the peak occurs in the −6 **NMZ** series at *m/z* 259.962; its normalized intensity and experimental error are 391.3 and −13.8 ppm, respectively. All peaks in the alkylbenzene series are thus accounted for up to *m/z* 399.9598. Whether the series extends beyond *m/z* 399.9598 can be determined by looking for the higher-mass homologs in the unassigned list. An examination shows that no additional members of the series are present.

Excluding the ions at *m/z* 77.9598 and 105.9598, which were used as references, the root mean square **(rms)** error over the whole −6 **SPZ** series through *m/z* 399.9598 and including the ions at *m/z* 119.9598 and 259.9598 is ±4.6 ppm. This error is typical of that observed for peaks in the entire spectrum as long as unresolved multiplets are not encountered.

Partial output from program **CH13CY** showing the ^{13}C isotope peaks of the hydrocarbons is given in Table 7. Although not nearly as intense as the ^{12}C peaks, the ^{13}C isotope peaks are prominent in the −8 and −10 **SPZ** series. At a resolving power of 40 000, the ^{12}C hydrocarbon fragment ions are resolved from the ^{13}C molecular ions to Mass 180. Comparison of the **CH^{13}C** and **CH** output show that the peaks are resolved—at least partially—to somewhat higher masses. For example, the *m/z* 216.9486/216.9531 doublet ($C_{15}{}^{13}CH_{24}$/$C_{16}H_{25}$) is resolved with intensities of 136.0/153.6, as shown in the −8 **SPZ** column of the second **CH13C** array and in the −7 **NMZ** column of the unassigned peaks at the end of the **CH** output (Table 2c). The expected intensity of the ^{13}C isotope peak is 166.7, which compares favorably with the observed intensity of 136.0.

Partial **CHS** output for the Cerro Negro fraction, given in Table 8, shows the presence of fragment and molecular ions containing one sulfur atom at low masses. At a resolving power of 40 000, these are expected to be resolved from the hydrocarbon ions to Mass 136. Comparison of the **CH** and **CHS** output (Table 2*b* versus Table 8) indicates that the C_3/SH_4 doublet is resolved to somewhat higher mass. For example, starting at Mass 127 the C_nH_{2n-13} peak is resolved, at least partially, from the $C_nH_{2n-3}S$ peak up to Mass 197 (rows containing *m/z* 183 and higher masses have been omitted from the bottom of Table 8). Because the maximum positive **SPZ** number for **CHS** homologs is +2 and because the difference in **SPZ** numbers for overlapping **CH/CHS** combinations is −10, the **CH** output will not contain overlapping peaks for **SPZ** numbers more positive than −8. Thus, the alkylbenzene series (**SPZ** = −6, Table 2*a*) will be free of overlap from **CHS** peaks.

The most abundant molecular ions for **CHS** homologs are found in the **SPZ** = −6 column (unedited normalized sum = 2554.9). These peaks are relatively free from overlap with **CH** peaks in the **SPZ** = −16 series (Table 2*b*) because the separation scheme limits the abundant hydrocarbon types to those containing one aromatic ring. By the same reasoning, relatively few **CH/CHS** overlapping peaks are expected for **SPZ** numbers more negative than −6 for **CHS** compound types.

Conclusion

High resolution mass spectra of mixtures consisting of components that can be classified into homologous series are readily processed by using Kendrick masses in a rectangular array laid out according to nominal mass Z series. This approach leads to partial assign-

TABLE 7—*Second array (partial) for $CH^{13}C$ combinations.*

NAME: RSPCNM. 8/12/86. CN 200-425 DEG C 1-RING COMPDS. RUN 1. SCAN NO.: 1
MAXIMUM PPM ERROR = 6.0 TOTAL ION CURRENT = 0.3152E+07 ELEMENTAL COMPOSITION: C(N−1)H(2N+Z)13C

NOMINAL MASS Z NUMBERS															
−11	−10	−9	−8	−7	−6	−5	−4	−3	−2	−1	0	1	2		
KENDRICK MASS DEFECT + 1.0000															
0.9218	0.9285	0.9352	0.9419	0.9486	0.9553	0.9620	0.9687	0.9754	0.9821	0.9888	0.9955	1.0022	1.0089		
SPECIFIC Z NUMBERS															
−12	−11	−10	−9	−8	−7	−6	−5	−4	−3	−2	−1	0	1	N	2N+1
MASS/INTENSITY/PPM ERROR															
								25.0	26.0	27.0	28.0	29.0	30.0	2	5
...	...	...	...	...	...	...	...	...	...	...	...	...	...	...	...
...	...	...	...	...	...	...	...	...	...	...	...	...	...		
...	...	...	..	...	...	...	...	...	...	...	...	...	...		
129.0	130.0	131.0	132.0	133.0	134.0	135.0	136.0	137.0	138.0	139.0	140.0	141.0	142.0	10	21
			36.4	37.1	79.3	62.4									
			−1.7	−3.3	−5.9	0.5									

143.0	144.0	145.0	146.0	147.0	148.0	149.0	150.0	151.0	152.0	153.0	154.0	155.0	156.0	11	23
	19.0	18.0	115.3		39.0	39.0									
	4.6	−0.3	−0.4		1.9	−4.5									
...	...	...	...	...	...	...	...	...	...	...	...	...	...	...	...
...	...	...	...	...	...	...	...	...	...	...	...	...	...		
...	...	...	...	...	...	...	...	...	...	...	...	...	...		
199.0	200.0	201.0	202.0	203.0	204.0	205.0	206.0	207.0	208.0	209.0	210.0	211.0	212.0	15	31
	72.5	105.3		168.2	38.3	48.5									
	3.2	2.1		−0.2	2.5	4.1									
213.0	214.0	215.0	216.0	217.0	218.0	219.0	220.0	221.0	222.0	223.0	224.0	225.0	226.0	16	33
33.1			52.7	136.0											
2.5			5.4	5.3											
227.0	228.0	229.0	230.0	231.0	232.0	233.0	234.0	235.0	236.0	237.0	238.0	239.0	240.0	17	35
20.0		48.6		243.0		60.2									
−3.5		4.3		−4.7		0.4									
241.0	242.0	243.0	244.0	245.0	246.0	247.0	248.0	249.0	250.0	251.0	252.0	253.0	254.0	18	37
		121.1													
		−4.0													
...	...	...	...	...	...	...	...	...	...	...	...	...	...	...	...
...	...	...	...	...	...	...	...	...	...	...	...	...	...		
...	...	...	...	...	...	...	...	...	...	...	...	...	...		
COLUMN SUMS															
829.4	708.2	1349.9	746.5	1570.8	156.6	404.2	9.7	0.0	0.0	0.0	0.0	0.0	0.0		

TABLE 8—*Second array (partial) for CHS combinations.*

NAME: RSPCNM. 8/12/86. CN 200-425 DEG C 1-RING COMPDS. RUN 1. SCAN NO.: 1
MAXIMUM PPM ERROR = 6.0 TOTAL ION CURRENT = 0.3152E+07 ELEMENTAL COMPOSITION: C(N)H(2N+Z)S

NOMINAL MASS Z NUMBERS															
−11	−10	−9	−8	−7	−6	−5	−4	−3	−2	−1	0	1	2		
KENDRICK MASS DEFECT + 1.0000															
0.8359	0.8426	0.8493	0.8560	0.8627	0.8694	0.8761	0.8828	0.8895	0.8962	0.9029	0.9096	0.9163	0.9230		
SPECIFIC Z NUMBERS															
−15	−14	−13	−12	−11	−10	−9	−8	−7	−6	−5	−4	−3	−2	N	2N−2
MASS/INTENSITY/PPM ERROR															
													44.0	1	0
				...	...	...	...	...	...	...	...	...	...	...	...
				...	...	...	...	...	...	...	...	...	...		
				...	...	...	...	...	...	...	...	...	...		
			104.0	105.0	106.0	107.0	108.0	109.0	110.0	111.0	112.0 58.4 0.1	113.0	114.0 26.8 −4.0	6	10
	116.0	117.0	118.0	119.0	120.0	121.0	122.0	123.0	124.0	125.0	126.0 79.8 0.2	127.0 37.4 −1.4	128.0 37.1 3.4	7	12
129.0	130.0	131.0	132.0	133.0	134.0	135.0	136.0	137.0	138.0 24.2 −3.4	139.0 157.2 −1.8	140.0 68.6 −2.1	141.0 39.8 3.2	142.0 21.1 −4.9	8	14
143.0	144.0	145.0	146.0	147.0	148.0	149.0	150.0	151.0 11.6 2.8	152.0 17.1 −1.8	153.0 143.4 1.6	154.0 22.4 3.3	155.0 97.5 3.5	156.0 11.0 −2.9	9	16
157.0	158.0	159.0	160.0	161.0	162.0	163.0 18.5 0.7	164.0	165.0 21.8 −0.5	166.0	167.0 180.0 −0.5	168.0	169.0 248.4 −1.3	170.0	10	18
...	...	...	...	...	...	...	...	...	...	...	...	...	...	...	...
...	...	...	...	...	...	...	...	...	...	...	...	...	...		
...	...	...	...	...	...	...	...	...	...	...	...	...	...		
COLUMN SUMS															
14.6	216.3	44.4	719.9	270.0	1807.9	203.3	1311.5	901.1	2554.9	1934.1	1767.2	2054.5	713.8		

ment of elemental compositions to masses in a spectrum that is easily edited to complete the assignments. The array output is especially useful for interpreting the spectra of complex petroleum fractions, and only a cursory examination of the output is required to identify the major compound types present in the mixture.

The fraction containing one-ring aromatic neutrals separated from the 200 to 425°C distillate of the Cerro Negro heavy petroleum was found to consist mostly of hydrocarbons with **SPZ** numbers ranging from -6 through -16. Homologs in the -8 and -10 **SPZ** series were the most abundant. Only traces of oxygen- and nitrogen-containing compounds were found, but sulfur compounds having formulas $C_nH_{2n+z}S$ with **SPZ** numbers ≤ -4 were found in significant quantities. These types of compounds are in agreement with those anticipated from the HPLC method used in the fractionation of the distillate.

Acknowledgment

Appreciation is expressed to the U.S. Department of Energy for support of this work under Cooperative Agreement DE-FC22-83FE60149 and through a fellowship administered by Associated Western Universities and the University of Oklahoma. Numerous helpful comments were provided by J. C. Brown, R. Frederick, J. B. Green, S. E. Scheppele, G. P. Sturm, Jr., S.-Y. Tang, and J. S. Thompson.

References

[*1*] Thomson, J., Reynolds, J., Yu, S. K-T., Tang, S.-Y., and Grigsby, R. D., "Effect of Degree of Aromaticity and Alkylation of Polyaromatic Hydrocarbons on Instrumental Response Factors," National Institute for Petroleum and Energy Research, Bartlesville, OK, 1986.

[*2*] Kendrick, E., *Analytical Chemistry,* Vol. 35, 1963, pp. 2146–2154.

[*3*] Scheppele, S. E., Chung, K. C., and Hwang, C. S., *International Journal of Mass Spectrometry and Ion Physics,* Vol. 49, 1983, pp. 143–178.

[*4*] Scheppele, S. E., Grindstaff, Q. G., Grigsby, R. D., McDonald, S. R., and Hwang, C. S., *International Journal of Mass Spectrometry and Ion Physics,* Vol. 49, 1983, pp. 179–209.

[*5*] Hamming, M. C. and Foster, N. G., *Interpretation of Mass Spectra of Organic Compounds,* Academic Press, New York, 1972, Chapter 7.

[*6*] Grigsby, R. D., Scheppele, S. E., Grindstaff, Q. G., Sturm, G. P., Jr., Taylor, L. C. E., Tudge, H., Wakefield, C., and Evans, S., *Analytical Chemistry,* Vol. 54, 1982, pp. 1108–1113.

[*7*] Clerc, R. J., Hood, A., and O'Neal, M. J., Jr., *Analytical Chemistry,* Vol. 27, 1955, pp. 868–875.

[*8*] Danti, A., "Report of Investigation of the American Petroleum Institute Research Project 44 and the Manufacturing Chemists Association Research Report," Carnegie Institute of Technology, Pittsburgh, 1960.

Author Index

Subject Index